Elite • 82

Samurai Heraldry

Stephen Turnbull · Illustrated by Angus McBride

Consultant editor Martin Windrow

First published in Great Britain in 2002 by Osprey Publishing,
Midland House, West Way, Botley, Oxford OX2 0PH, UK
44-02 23rd St, Suite 219, Long Island City, NY 11101, USA
E-mail: info@ospreypublishing.com

Transferred to digital print on demand 2010

First published 2002
5th impression 2009

Printed and bound by PrintOnDemand-Worldwide.com, Peterborough, UK

A CIP catalogue record for this book is available from the British Library

ISBN: 978 1 84176 304 0

Series Editor: Martin Windrow

Design by Alan Hamp
Index by Alan Rutter
Originated by Magnet Harlequin, Uxbridge, UK

Dedication
To our daughter-in-law Louise Turnbull

Author's Note
This book is intended to provide a comprehensive guide to the art and science of heraldry in Japan at the time of the samurai.
As it is intended for military enthusiasts I have kept to a minimum any descriptions of heraldry as used purely for decoration. The
emphasis is on the use of heraldry in war from about the time of the Gempei Wars of the 12th century to the Shimabara Rebellion
of 1638. (It is also for this reason that I have added a final section on creating flags for an authentically modelled wargame army.)

As space is limited I have kept brief any biographical notes on the named samurai who used the flags and the motifs which follow;
I refer the interested reader to my book *The Samurai Sourcebook* (Cassell, 1998) for further details. I have also tried to avoid
duplicating illustrations of any families that I have covered in previous books; so, for example, the heraldry of the chief retainers of
Takeda Shingen and Uesugi Kenshin can be found in my *Samurai: The Warrior Tradition* (Cassell, 1996).

The study of samurai heraldry is constantly uncovering new material. Interested readers can visit the author's website at
www.stephenturnbull.com for a regular update on new discoveries and useful links free of charge. Any new material which
readers send in will be included so that all can share it.

Acknowledgements
I would like to thank Mike O'Brien for providing some very useful ideas early on in the project; Noboru Koyama of Cambridge
University Library for allowing me to examine their hand-painted copy of Kyuan's *O Uma Jirushi*; the Sonkei Kaku Library in Tokyo
for supplying a photograph of the Suemori scroll; the Hikone Museum for details of the Osaka Screen; Brian Bradford for pictures
of Korean flags; and Philip Abbot of the Library of the Royal Armouries, Leeds, for books of *mon*. All the other material was
collected by the author 'on location' in Japan.

Artist's Note
Readers may care to note that the original paintings from which the colour plates in this book were prepared are available for
private sale. All reproduction copyright whatsoever is retained by the Publishers. All enquiries should be addressed to:
Scorpio Gallery, PO Box 475, Hailsham, E.Sussex BN27 2SL, UK
The Publishers regret that they can enter into no correspondence upon this matter.

FOR A CATALOGUE OF ALL BOOKS PUBLISHED BY OSPREY
MILITARY AND AVIATION PLEASE CONTACT:

Osprey Direct, c/o Random House Distribution Center,
400 Hahn Road, Westminster, MD 21157
Email: uscustomerservice@ospreypublishing.com

Osprey Direct, The Book Service Ltd, Distribution Centre,
Colchester Road, Frating Green, Colchester, Essex, CO7 7DW
E-mail: customerservice@ospreypublishing.com

www.ospreypublishing.com

SAMURAI HERALDRY

THE FUNCTIONS OF HERALDRY

Through the art and science of heraldry the armies of different ages and different regions around the world have been able to distinguish friend from foe in the confusion of battle. The practices employed have ranged from the painting of simple patterns or devices on ancient warrior shields, through the embroidery of complex designs on associated sequences of regimental flags, and back to simplified divisional symbols painted on the hulls of modern tanks. Some form of heraldry is found in nearly every military society.

As well as merely identifying allies and opponents, the practice of heraldry has also performed the function of glorifying a particular individual by ensuring that his personal achievements were easily recognised. The extension of this has been the introduction of a hereditary element into heraldry, so that a warrior's descendants would always be associated with the exploits of a particular brave ancestor. In this sense, heraldry has rightly been called 'the shorthand of history'.

All these aspects of heraldry applied as readily to Japanese as to European military history. For many centuries Japan was a military society, and five hundred years of civil wars made the ability to distinguish at a glance allied armies from rival contingents a paramount necessity. The importance placed upon personal identification and display was also very great in a warrior society that prized individual achievement above all else – even, it would appear from some accounts, above the need to actually win battles. Even in the massive and well-organised encounters of the 16th century a samurai's personal prowess was still highly valued, and achievements such as being the first into battle or taking a noble opponent's head were as eagerly pursued as in a previous age.

In both these honourable examples the use of heraldic devices was a vital aid towards establishing the truth of a samurai's claim. During a siege the presence of a warrior's flags on an enemy castle wall proved who had been the first to fight his way in. During the head presentation ceremony that invariably followed a battle, two separate forms of heraldic

A suit of armour owned by Toyotomi Hideyoshi, which makes considerable use of *mon* for decorative purposes. There is also a quasi-heraldic element in the shape of a sun disc – *hi no maru* – picked out in red on blue from the silk lacing of the breastplate.

The earliest use of devices identifiable as *mon* are to be found in connection with the Japanese imperial family. In this illustration from the *Heiji Monogatari Emaki* the emblem of 'nine stars' appears on the side of an imperial ox-cart used during the Heiji Rebellion of 1160. Note that this is the only 'heraldry' visible. The samurai wear no personal heraldic identification, and would fight under a distinctive banner instead.

identification aided the recognition of personal exploits. One was the identification of the victim, whose severed head was presented to the victorious samurai's lord as an invoice for reward. The other was confirmation by eyewitnesses that the particular samurai had actually killed the man whose head he was presenting. In both cases the heraldic flag or other device associated with either party was vital evidence.

Apart from this use of heraldry for the benefit of an individual in a samurai army, flags and colours were also a vital element in battlefield organisation. By the use of different coloured heraldic flags the separate units of an army could be identified and controlled. In addition, the prominent heraldic standards used by generals – called *uma jirushi* or 'horse insignia' – provided a rallying point on the battlefield[1].

1 In the body text Japanese terms are only italicised at their first use; in the captions, only nouns referring to types of heraldic display and charge are italicised throughout.

GLOSSARY

Hata bugyo Officer responsible for organisation, ordering and verifying of an army's flags.
Hata jirushi Tall, narrow flag flown vertically from crossbar strung by cords to top of vertical shaft.
Hata sashi Flag-carrier, standard-bearer.
Hi no maru Sun disc emblem, displayed as heraldic charge.
Horo Cloak-like, bag-shaped cloth worn over light framework behind rider's back, to identify a leader's messengers, aides or bodyguards.
Ibaku (also *maku*) Curtains used to screen off a leader's headquarters position in the field.
Kami God/goddess of the Shinto religious pantheon.
Kasa jirushi Very small identification flag of army or unit, attached to helmet.
Maku See *ibaku*.
Mon Emblem identifying individual or family, displayed in his/their heraldry.
Nobori Tall, narrow flag with two edges attached to both a vertical shaft and a short upper crossbar forming an inverted L-shape.
Sashimono Emblem, of varying size and form but often a tall, narrow flag, attached to rear of cuirass to identify army or unit.
Sode jirushi Very small identification flag of army or unit, attached to shoulder plates of armour.
Tsukai ban Leader's unit of messengers/aides.
Uma jirushi Leader's large standard for battlefield identification/location; either a conventional or *nobori*-shaped flag, or a three-dimensional emblem. *O uma jirushi* = 'great', *ko uma jirushi* = 'lesser' standard.

Four centuries later Japanese heraldry had developed into a complex practice, if never a 'code' in the European sense. This is the full heraldic display of Arima Toyouji (1570–1642), a *daimyo* or baron who fought at Osaka (1614–15) and Shimabara (1638) – the final battles of the 'samurai age' proper before Japan settled into the centuries-long peace of the Edo Period under the Tokugawa shoguns. All the flags are in black and white, and all the three-dimensional devices are gold lacquered. (1) Double back-flag worn by Arima's *ashigaru* or common footsoldiers; (2) three-dimensional sunburst back-device worn by his messengers; (3) three-dimensional crescent *sashimono* back-device worn by his *samurai* warriors; (4) Arima's 'lesser standard', an elaborate three-dimensional gold trefoil; (5) his 'great standard' bearing his *mon* or family emblem, which is repeated on (6), the *nobori* flags carried by his attendants and troops.

THE USE OF THE *MON*

The best known features of Japanese heraldry are the devices called *mon*, which are usually simple yet elegant motifs based on plants, heavenly bodies, geometric shapes or, more rarely, animals. By their very simplicity mon are much more easily recognisable than European coats-of-arms, even if the European system of quartering and labels provides a more precise identification of an individual. Unlike a European blazon, however, the particular colour of a mon was never specified. They are usually depicted as black upon white or another light colour, or in white upon black or another dark coloured field. Particular colours were introduced in the design of the flags upon which the mon were most often displayed on the battlefield.

The direct parallel to mon in European heraldry is the badge, which was sewn on to a soldier's jacket and used for the same purpose of quick recognition; but as mon also served as the equivalent of the coat-of-arms, the layman's understanding of them as 'Japanese family crests' is not too wide of the mark. Mon often appear as the actual crest above the peak of a Japanese samurai's helmet, and are also seen on scabbard designs, on flags at shrines, or as a purely decorative element on clothing.

It should be noted that, unlike the quartering on a European knight's surcoat, mon did not always appear on the sleeveless *jinbaori*, which was the Japanese equivalent of the surcoat. This garment, which was used only by high-ranking samurai, and rarely while actually fighting, tended to be lavishly embroidered with designs that had no particular heraldic significance.

Most of the illustrations in this book involve mon, but it is important to stress at the outset that mon were only part of the story of Japanese heraldry, as the following pages will show. Mon did not always appear on battle flags, because bands of contrasting colour sometimes provided all the identification that was necessary. From the 16th century onwards the overall layout, colour and design of flags, from the large ones carried as standards to the smaller ones worn on the back of suits of armour, were

(1) The *kiku mon*, the Japanese imperial chrysanthemum, which probably owes its origins to a stylised representation of the sun's rays.

(2) The *kiri* (paulownia) *mon* of the Ashikaga family, bestowed upon them originally by a grateful emperor, and used in various combinations with other designs as described in the text.

(3) The *kikyo* (Chinese bellflower) *mon* of Ota Dokan (1432–86), one of the first 'Sengoku daimyo' – the provincial barons who emerged as powerful regional warlords following the political and social upheavals of the Onin War (1467–77).

(4) The *ume* (plum blossom) *mon* of Tsutsui Junkei (1549–1641), who became notorious for his reluctance to join in the battle of Yamazaki in 1582 until it became clear which side would win.

(5) The *katabami* (oxalis) *mon* of Chosokabe Motochika (1539–99), who took over the entire island of Shikoku before being defeated by Hideyoshi in 1585.

(6) The *tsuru* (crane) *mon* of Mori Nagayoshi (1558–84), who was killed at the battle of Nagakute. This is identical to the device used by Nanbu Toshinao, son of Nobunao (1546–99).

prescribed as carefully as the mon, and therefore add another dimension to the study of samurai heraldry. Nevertheless, it was largely through his mon that an individual samurai and his family were known, regardless of what other symbols and designs they might have displayed in actual combat over the centuries.

Japanese Imperial heraldry

The earliest mention of flags in the context of Japan is found in an ancient Chinese chronicle, and refers to messengers being sent to Japan from China bearing yellow banners, yellow being the most highly esteemed colour. The early emperors of Japan are also recorded as giving flags of gold brocade along with presentation swords as gifts to the warriors whom they commissioned and sent out to chastise rebels against the throne.

The emperors of Japan themselves are traditionally associated with the *kiku no go mon*, the well-known design of a 16-petalled chrysanthemum. Its origin is obscure, and it has been suggested that it may be derived from an image of the sun rather than a flower, with what appear to be flower petals actually being the sun's rays. Whatever its origin, the kiku no go mon was an emblem which was reserved for imperial use alone; and in 1871, following the Meiji Restoration which led to the establishment of modern Japan, a decree was issued forbidding anyone to use any mon that could be mistaken for the imperial symbol. Regulations for the display of the imperial chrysanthemum were also set out at the same time, and specified that the 16-petalled flower was for the reigning emperor only, with the small circle in the middle indicating that it was being viewed from the front. Imperial princes had a 14-petalled flower, with the calyx in the centre indicating that it was being viewed from below. Emperor Go Daigo, whose abortive attempt to restore imperial power in the 14th century led to the Nanbokucho Wars, used a 17-petalled chrysanthemum during his years of exile.

The earliest illustrations of mon in Japanese history are associated with emperors and their courts. They may be found in scrolls depicting imperial processions, where various simple designs are seen as early as the Nara Period (AD 710–784) on the sides of the ox-carts used for transporting imperial officials. A favourite device was the 'nine stars', which consisted of one disc surrounded by eight others, but at this period the mon had no military significance.

EARLY HERALDRY

From the 10th century AD onwards the powerful landowning families of Japan began a struggle for supremacy that was eventually to lead to the curtailment of imperial power and the relegation of the emperor to the position of a figurehead. The warriors who served these lords were called

The doves of Hachiman, the tutelary deity of the Minamoto, appear in the foreground of this section of the *Gosannen Kassen Emaki* scroll on the *maku* (field curtain) of Minamoto Yoshiie during the Later Three Years' War of the 1080s AD. The use of this design as an actual heraldic device may be a later attribution.

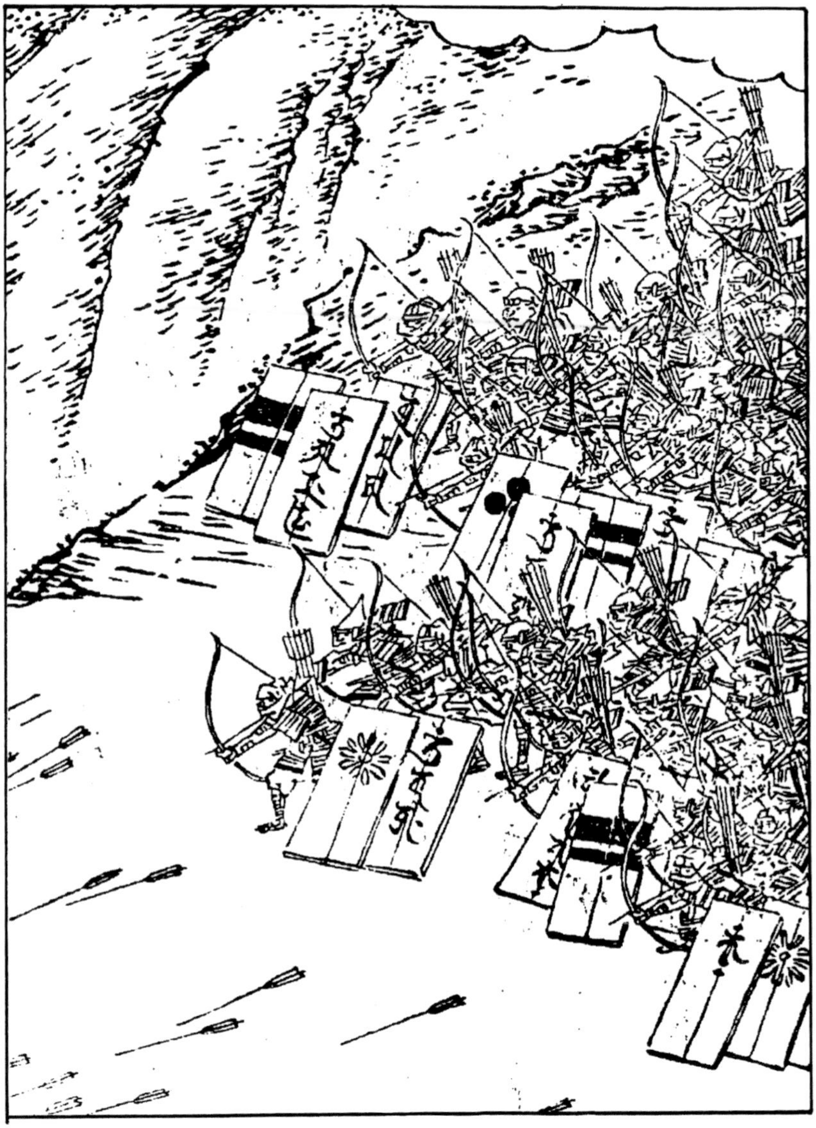

Various *mon* and black bands were painted on footsoldiers' wooden mantlet shields to distinguish families and units in battle. During the 12th-century Gempei Wars this was one of three locations where *mon* would appear, the others being flags and field curtains. In this illustration the shields belong to an army of *sohei* (soldier monks), so Buddhist slogans written in *bonji* – Sanskrit characters – also appear.

The crescent moon and star flag of the Chiba, which alludes to the help supposedly rendered to their ancestors in battle by Myomi Daibosatsu, a deity associated with the constellation of the Plough.

samurai. Two families in particular, the Taira and the Minamoto, fought a long series of wars that ended with the destruction of the Taira and the founding by the Minamoto of the institution of government by the *shogun,* Japan's first military dictatorship, in 1192.

The long rivalry between the Taira and the Minamoto clans provides an early and well-recorded example of the use of heraldry in Japan. The military operations carried out during the Hogen Rebellion of 1156, the Heiji Rebellion of 1160 and the Gempei War of 1180–85 were all characterised by the use of flags for army identification and organisation. The particular type of flags used at this time were called *hata jirushi,* and consisted of long 'streamers' hung from a horizontal crossbar strung to the top of a tall pole. The hata jirushi were carried into battle by footsoldiers, or by mounted men who rode alongside the elite samurai mounted archers.

The one aspect of heraldic display at this time that is not seriously disputed is the use of red flags by the Taira and white flags by the

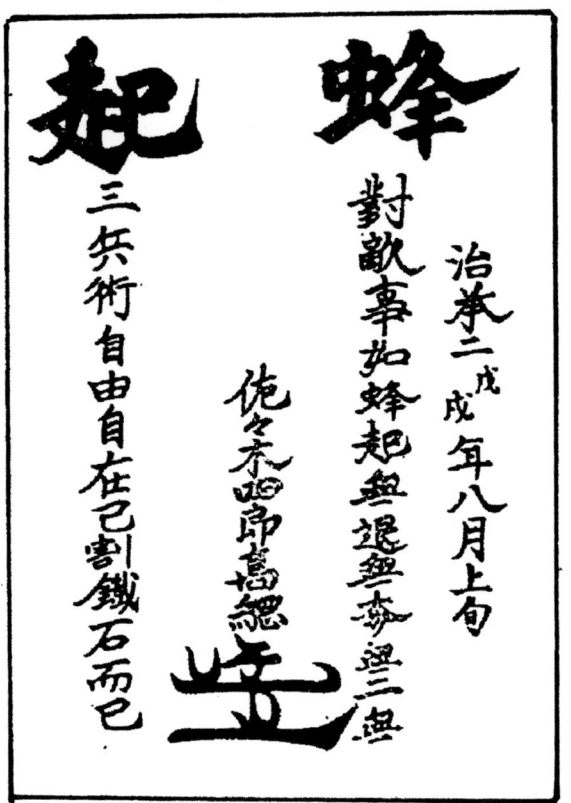

The banner with a prayer to the war god Hachiman used by Sasaki Takatsuna, the hero of the second battle of the Uji in 1184, when he led the crossing of the river.

Minamoto. There are many references to these colours in the contemporary literature and in the *gunkimono* or 'war tales' written about these campaigns over the next two centuries. At the battle of Dan no Ura in 1185, we are told, the sea ran red with the blood from the slain Taira samurai and the red dye from their banners.

As for the designs of the mon attributed to the two families, there is a very long tradition that associates the Taira with the mon of a stylised black butterfly, and the Minamoto with a design based on the *rindo* (gentian) plant – see Plate A4 & A3. Unfortunately there is no unequivocal proof for the actual use of these two mon during the 12th century, but every later painted screen and woodblock print depicting the Gempei War invariably shows them on the red or white flags. In the shrine on the site of the battle of Yashima (1184) is displayed a banner said to have been used by the Minamoto during the battle. It is plain white, and appears to have had no emblem added to it.

So did the Taira and the Minamoto use mon on their flags? In the 18th-century work entitled the *Honcho Gunkiko* the historian Arai Hakuseki (1656–1725) quotes a passage from the *Heike Monogatari* referring to the Taira's red flags and the Minamoto white flags, and then goes on to say that the *ibaku* (curtains) had their respective mon on them. The ibaku, which were also called *maku*, were another characteristic feature of the samurai battlefield. They were large cloth curtains suspended from ropes slung between poles to screen off a private area for the commander and his closest retinue. These curtains were such an established feature of samurai life that when Minamoto Yoritomo set up his shogunate in 1192 he called the institution the *bakufu*, 'the government behind the curtain'.

During the 16th century it was usual for large versions of the commander's mon to be embroidered or printed on the maku. It is, however, by no means clear whether any designs appeared on the Taira and Minamoto maku during the 12th-century Gempei Wars. Illustrations of the Gosannen ('Later Three Years') War, fought by Minamoto Yoshiie from 1086 to 1089, show a device of birds on Yoshiie's maku, an allusion to an incident when he was warned of an ambush by wild birds rising in disorder from a field; but this may well be an instance of artistic licence by the scroll illustrator, who has tried to include the famous legend in some way.

There is, however, considerable circumstantial evidence that some form of device, either the actual Taira and Minamoto mon or other designs, did appear on their flags on many occasions. For example, at one stage during the Gempei War two branches of the Minamoto were in action in different areas of Japan, and eventually fought each other. In 1184 Minamoto Yoshitsune was sent to quell the disturbances caused in Kyoto by his boisterous cousin Minamoto Yoshinaka, who had defeated the Taira the previous year and occupied the capital with much looting and destruction. In the *Heike Monogatari* account there is a description of the panic caused in Kyoto when an army with white banners was seen approaching. The citizens were afraid that Yoshinaka was coming to pillage them again, but the Minamoto samurai were soon identified as Yoshitsune's men because their 'insignia were different', and calm was restored. In another section of the *Heike Monogatari* a group of Taira partisans are mentioned as having 'a badge of oak leaves'. Both these passages imply that various mon were added to the white and red flags; and in a further passage such an alternative mon is described, when a member of the Kodama family, who were allied with the Minamoto, is recorded as using a white flag with the mon of a fan. It therefore seems reasonable to conclude that both Taira and Minamoto mon were often displayed upon their flags, and that these were probably the designs of the butterfly and the gentian.

Another place where motifs might appear during this time was on the front of the large wooden mantlet shields which the footsoldiers used for protection on the battlefield. As well as mon, black bands sometimes appear to have been painted on the front. Similar black bands also appear on the white and red flags, probably to distinguish units within an army, or perhaps to show cadet branches of the same family.

Mon were not the only devices used on flags, curtains and shields during the Gempei War. There are several authentic examples of flags bearing religious or personal inscriptions written in black ink. These include Buddhist prayers, poetic exhortations to bravery, and invocations of various Shinto *kami* (gods) to whom a warrior was particularly devoted. Sasaki Takatsuna, the hero of the crossing of the Uji River in 1184, is supposed to have used a white rectangular banner with a prayer written on it in black ink and dated 'Jisho 2' (1178).

Heraldry and samurai heroes

The problem over the attribution of the butterfly and gentian mon to the Taira and Minamoto is but one example of the difficulties that arise

OPPOSITE **The banner of Kumagai Naozane, alluding to the escape of Minamoto Yoritomo in 1180 described in the text. The symbolism is considerable. There are two doves in the middle, and at the top two dove shapes form *ha*, the first character of 'Hachiman'; all refer to the doves whose disturbed flight from the tree where he was hiding distracted Yoritomo's searching enemies. The snowflake motif below reflects the fact that the incident occurred in wintertime.**

In this section from the *Moko Shurai Ekotoba* (Mongol Invasion Scroll) depicting the events of 1274, a *hata jirushi* with a prominent *mon* of two feathers (left) is paraded in front of a Japanese commander.

when a particular mon is credited to a famous family or a celebrated samurai hero. It was quite common for a particular family to adopt a mon or other design because of an ancestor's brave exploit, the choice of motif referring in some way to this proud incident in family history. As in the case noted above of Minamoto Yoshiie being credited with the design of flying birds on his maku, it is often impossible to determine whether or not the device was actually used by the named warrior, or retrospectively attributed to him by proud descendants who themselves made use of it (this is perhaps comparable to the practice known in European heraldry as 'canting arms').

A case in point is the Chiba family. The *Taiheiki,* which describes the Nanbokucho Wars of the 14th century, refers to the use by the Chiba family of two particular mon: a pattern of either seven or nine stars represented by discs, and a very distinctive star and crescent. The interesting point about these mon is that their adoption was linked by the family to an incident that happened in 931, when a Chiba army were on the point of annihilation during a battle with their rivals. The constellation of the Plough was very prominent in the night sky over the battlefield, so the Chiba prayed to the deity Myomi Bosatsu, who was associated with the seven stars of the Plough. As a result Myomi Bosatsu appeared to them in a vision, and the Chiba were saved. We do not know for certain whether the Chiba used the mon in the immediate aftermath of their salvation; nor is there any absolute proof that they used it during the Gempei War, when Chiba Tsunetane (1118–1201) supported Minamoto Yoritomo, although this would seem likely. Instead, the mid-14th century provides the first documentary evidence of its actual display, and a later Chiba used the device in 1455.

Kumagai Naozane, who was one of the legendary heroes of the Gempei War, provides another example of these puzzles over attribution. Naozane first served the Taira, but then secretly switched his allegiance to Minamoto Yoritomo. When Yoritomo was on the run after his defeat at Ishibashiyama in 1180, Naozane was sent to track him down. He found Yoritomo hiding in a tree and, not wishing to betray his new master, stuck his bow into the branches and disturbed two doves. The doves flew out, and their presence in the tree satisfied his companions that no one else could have been hiding there. The incident is given added poignancy by the fact that doves are regarded as the messengers of Hachiman Dai Bosatsu, the deified Emperor Ojin, who was both the tutelary kami of the Minamoto and, effectively, the Japanese god of war.

On the banner later attributed to Kumagai Naozane there appear two doves and the name of Hachiman, and in one version the first character *ha* is written using two dove shapes. There is also the mon of a snowflake, which refers to the fact that the search for the fleeing Minamoto Yoritomo took place in wintertime. It may well be that Naozane used this banner during his later career, which included the battle of Ichinotani in 1184; but there is no independent confirmation to back up this attractive heraldic legend. (Strange to relate, other versions of the tale attribute Yoritomo's salvation to an entirely different person: Kajiwara Kagetoki.)

Heraldry and the Mongol invasions

The Minamoto shoguns were usurped in their turn in 1219 by the Hojo family, who ruled from the city of Kamakura under their distinctive mon

of three triangles arranged in a pattern like fish scales. This design is usually referred to as being based on actual fish scales (cf Plate C1), but Hojo family tradition describes the scales as those of a dragon. The goddess Benten had appeared to Hojo Tokimasa in the 12th century on the island of Enoshima, and when she entered the sea he noticed that she had a dragon's tail and had left some scales behind on the beach.

Overall, the 13th century was a comparatively peaceful period for Japan, disturbed only by a short-lived attempt at imperial restoration, and by the dramatic defeat of the Mongols' attempted invasion. The practice of displaying mon on hata jirushi flags continued throughout this time, and in 1219 a member of the Iriki family of Satsuma is noted as having mistletoe as his mon; however, pictorial sources are non-existent before the appearance of the *Moko Shurai Ekotoba* (Mongol Invasion Scroll), one of the most important contemporary sources for the appearance of samurai armies.

This famous scroll was painted under the direction of a certain Takezaki Suenaga so that he could use it as a basis for pressing his claim for reward for fighting against the Mongol army in 1274. Apart from valuable illustrations of costume and equipment in a narrowly datable context, it also includes several excellent depictions of heraldry in the latter part of the 13th century. The Mongol Invasion Scroll shows us that heraldic identification was still provided only by flags carried by a samurai's attendant, either a mounted man or a footsoldier. All the flags depicted on this scroll are of hata jirushi form.

Most of these flags are white banners differentiated from one another by the use of mon, which are clearly linked in the accompanying text to individual samurai, thus providing concrete proof that the identification of individuals by the use of mon was well established by the 1270s. For

OPPOSITE **A *hata jirushi* bearing a *mon*, and another plain white example, are carried during one of the 'little ship raids' against the invading Mongol fleet.**

example, in the section showing a march-past by the Japanese army there is a white hata jirushi bearing a black mon with a double feather design. In another section a footsoldier is shown running along beneath a shower of arrows, carrying a hata jirushi identified on the scroll as being the personal flag of Takezaki Suenaga; this flag is of two colours, white in the upper two thirds and dark blue in the lower third, with a black mon in the white section consisting of three open lozenges above the Japanese character *yoshi* meaning 'old'. The Shoni family also appear on the Mongol Invasion Scroll leading one of the 'little ship raids' on the Mongol fleet. Their banner is a white hata jirushi with a mon of four open black lozenges arranged in a lozenge pattern.

14th–15th CENTURY HERALDRY

The defeat of the Mongols, although a glorious episode in samurai history, gravely weakened the resources of the Hojo family. In 1331 Emperor Go Daigo led a rebellion in an attempt to overthrow the Hojo and restore the power of the emperors to what it had been before the Gempei War. In this he was unsuccessful, and the result was a series of conflicts between two rival lines of 'northern' and 'southern' emperors – the Nanbokucho or 'Wars between the Courts'.

Emperor Go Daigo was assisted in his endeavours by a number of famous samurai such as Nitta Yoshisada, who captured Kamakura in 1333; but few names are more celebrated in Japanese history than Go Daigo's intrepid follower Kusunoki Masashige, whose loyalty to the emperor led to his death at the battle of Minatogawa in 1336. This was a battle that Kusunoki did not want to fight, but the emperor turned down his suggestion of a defensive war, so Kusunoki went willingly to his death.

Kusunoki's suicide at Minatogawa made him the paragon of imperial loyalty, and this is reflected in his heraldry. Numerous scroll paintings and screens survive to illustrate his loyal exploits, and in most cases his long hata jirushi banner is shown as a white flag on which is depicted a version of his famous *kikusui* (chrysanthemum on the water) mon – see Plate B1. The device was given to him by the emperor whom he served so faithfully, and represents the imperial mon of the chrysanthemum being kept afloat by the Kusunoki. This use of the imperial mon, even in a modified form, by a samurai who was not a member of the imperial family is unique in Japanese history, and shows the great esteem in which Kusunoki Masashige was held. In fact, during the Meiji Period at the end of the 19th century which followed the modern restoration of imperial power, Kusunoki Masashige was celebrated as a symbol of imperial loyalty and as an example to the nation.

There are several variations of the floating chrysanthemum design among Kusunoki's flags that have been preserved in temples and in contemporary paintings. In addition to the kikusui mon, most (see Plate B2) have the five Chinese characters *hi, ri, ho, ken* and *ten*, which is explained as follows. *Hi* (error) represents the principle of unreasonableness; *ri* is justice, *ho* is the law, and *ken* is authority. *Ten* stands for the way of heaven, and of all these it is only through following the way of heaven that one may gain victory.

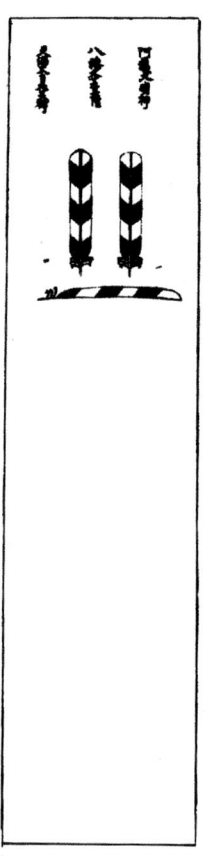

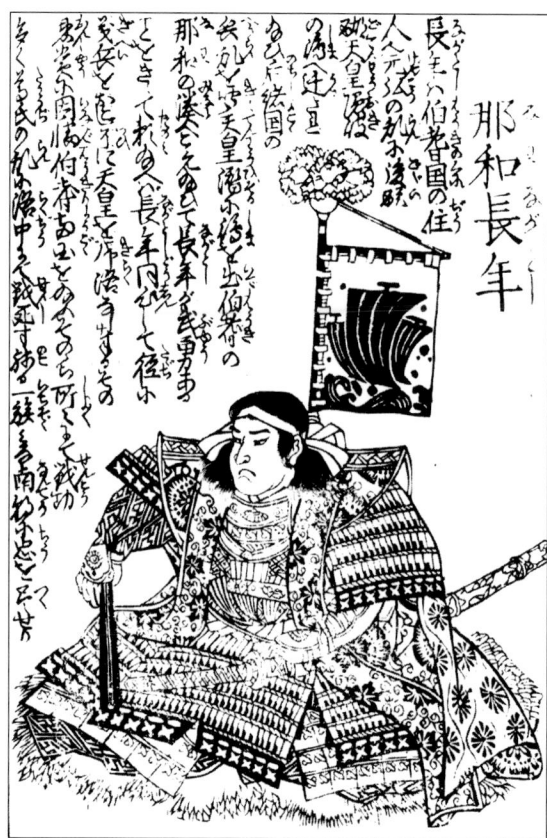

FAR LEFT **An early example of a *mon* based on a design of feathers appears on this white *hata jirushi* of the Kikuchi family of Kyushu, who fought for Emperor Go Daigo in the Nanbokucho Wars of the 14th century. Kikuchi Taketoki (1293–1334) was one of their most illustrious members.**

LEFT **Nawa Nagatoshi (died 1336) is shown here with the 'ship on the water' *mon* indicating that his loyalty to his lord Kusunoki Masashige was as firm as that of Kusunoki to the emperor. The later artist has, however, committed an error in giving Nawa a *sashimono* – this back-flag was not introduced until the 16th century. Nagatoshi's *mon* would actually have been carried on a *hata jirushi*.**

A similar principle of supporting one's lord whatever the consequences is expressed through the mon of Nawa Nagatoshi, a follower of Kusunoki Masashige who died during the battle for Kyoto in 1336. He used the design of a ship on his flags. Alluding to Kusunoki's own design of a chrysanthemum floating on the water, Nawa's mon was supposed to represent the lord (the ship) being supported by his retainer (the water), just as Kusunoki (the water) supported his lord the emperor (the floating chrysanthemum).

At about the same time as Kusunoki's exploits we also see further examples of slogans written in ink on banners and used on battlefields. Asuke Jiro, who was active early in the Nanbokucho Wars, provides one of the best examples of a slogan being written on a banner in all Japanese history, although in his case it is so long it is almost an autobiography, and resembles the proclamations of pedigree traditionally shouted on a battlefield before commencing single combat with a worthy opponent. The full inscription translates as follows:

'I was born into a warrior family, and loved courage like the youth of ancient times. My military strength and determination were such that I could cut a fierce tiger to pieces. I studied the way of the bow, and learned well the techniques of warfare. Being graciously subject to the lord of heaven, when face to face on the battlefield my desire was for a decisive encounter. At the age of 31 while having an attack of fever I went to Oyama and ran an important enemy through, holding my loyal exploits in high regard, and not partaking in immorality. My name will be praised throughout the whole world and bequeathed to my descendants as a

OPPOSITE **The banner of Kojima Bingo Saburo Takatoku, which bears a further example of a long invocation of the gods being used as a heraldic device.**

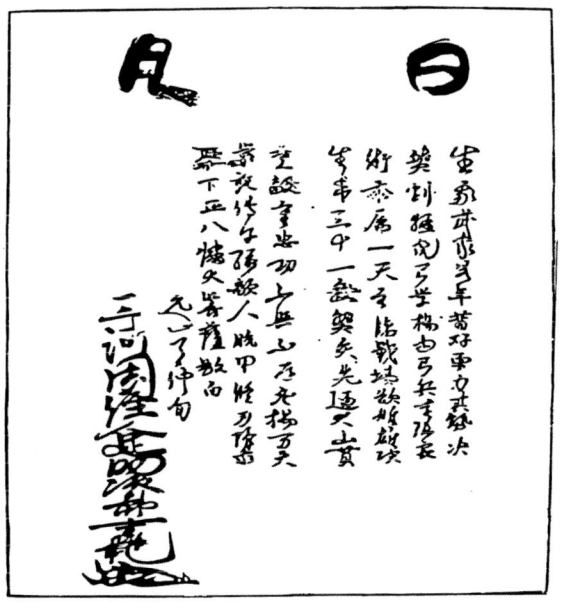

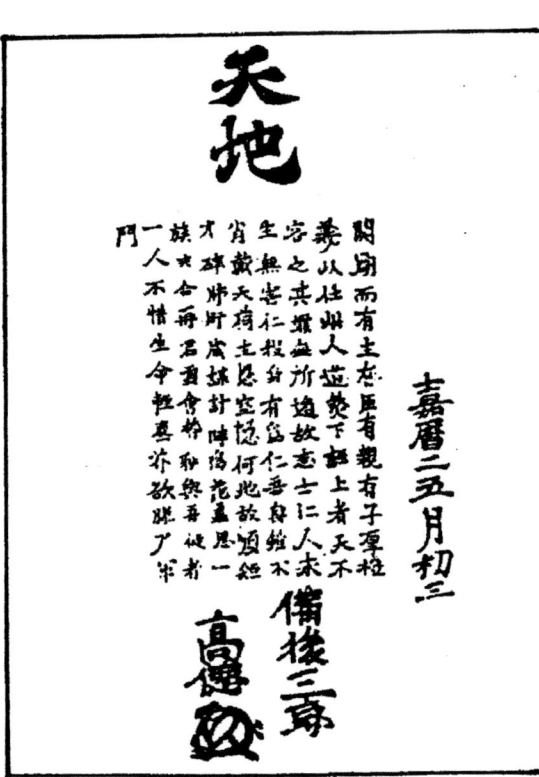

ABOVE **The banner carried in the Nanbokucho Wars for Asuke Jiro, which gives his full autobiography written in black ink on white – for a translation see the body text.**

glorious flower. Enemies strip off their armour and surrender as my vassals, I who have mastered the sword. The righteous Hachiman Dai Bosatsu. Sincerely, Asuke Jiro of Mikawa province.'

Another example of an elaborate inscription occurs on the flag of Kojima Bingo Saburo Takatoku. In this case it is written in Chinese and signed by Takatoku himself with the date 1327.

The Ashikaga shoguns

As well as leading to half a century of rivalry between 'northern' and 'southern' emperors, Go Daigo's failure also resulted in a new dynasty of shoguns taking power. Ashikaga Takauji (1305–58) became the first Ashikaga shogun in 1338, and his descendants supplied shoguns for the next two centuries. The Ashikaga were descended from the Minamoto – a necessary qualification for an aspiring shogun – and one way of proclaiming their legitimacy was by using as a heraldic device on their flags an invocation of Hachiman Dai Bosatsu, the tutelary deity of the Minamoto. This was a very ancient design, and had been adopted by the first samurai to take the name of Ashikaga. He was Minamoto Yoshikuni (d.1155), the son of Minamoto Yoshiie (1041–1108), who settled at Ashikaga in Shimotsuke province and took the name from the place.

Invocations of Hachiman were not restricted to the Ashikaga, of course, and they also appear on the flags of the contemporary Akamatsu family, who played a prominent part in Japanese military and political life from the mid-14th century onwards. Akamatsu Enshin Norimura (1277–1350) used a simple hata jirushi with the inscription 'Hachiman Dai Bosatsu' over the character *matsu*. In this design the first character of the name of Hachiman – *ha* – is represented pictorially by two doves, just as in the earlier example of Kumagai Naozane noted above. Enshin fought first for Go Daigo, but when he was stripped of his rewards he passed into the service of the Ashikaga, and fought at Minatogawa in 1336. Akamatsu Norisuke (1312–71), the son of Enshin, used a banner with a prayer to Hachiman written in black ink and dated 'Genko 2' (1332).

One particular heraldic design was, however, initially restricted to the Ashikaga family. This was the *kiri* mon, a stylised version of the paulownia plant, which was originally an imperial crest second in importance only to the full chrysanthemum design. The Ashikaga always supported the northern court, and the kiri mon was originally bestowed upon them by the first 'northern emperor' whom they served. As the years went by

Akamatsu Enshin Norimura (1277–1350) used a simple *hata jirushi* with the inscription 'Hachiman Dai Bosatsu' over the character *matsu*. The first character of 'Hachiman', *ha*, is once again represented pictorially by two doves.

the use of the kiri mon came to stand for imperial acceptance and commission, and other generals began to adopt it following the fall of the Ashikaga in 1568 (the most prominent in this regard being Toyotomi Hideyoshi, the late 16th-century unifier of Japan whose humble origins prevented him from becoming shogun). During the 14th–15th centuries, however, only the Ashikaga used the kiri mon, which usually appeared alone in black on a white banner. The design on the banner of the last Ashikaga shogun, Ashikaga Yoshiaki (1537–97), appears to be the sole exception; he dispensed with the kiri and instead combined a red sun disc with the ancient Ashikaga device of an evocation of Hachiman (see Plate B4).

The Kamakura branch of the Ashikaga family, who were descended from a younger son of Ashikaga Takauji and therefore did not provide any of the shogunal line, made use of similar devices to those of the senior branch. For example, Ashikaga Shigeuji (1434–97) fought a long civil war in the Kanto area with Ashikaga Masatomo (1436–91), brother of the shogun Ashikaga Yoshimasa (1435–90). Shigeuji was based at the town of Koga, and was styled Koga-kubo ('governor at Koga'). Their long conflict coincided with the better known Onin War in Kyoto. Ashikaga Shigeuji's banner was a striking hata jirushi which bore both the kiri mon and a sun disc (see Plate B3).

In addition to the examples noted above that are associated with the shogun's family, other families continued to use ancient mon bestowed upon their ancestors, and once again the Chiba provide the best example. Chiba Tanenao, who died in 1455, supported the Koga-kubo Ashikaga Shigeuji, and when the latter was defeated he committed suicide along with his son. Tanenao's banner displayed the mon of a star and crescent, which associated the family with their ancestor's salvation over five centuries before.

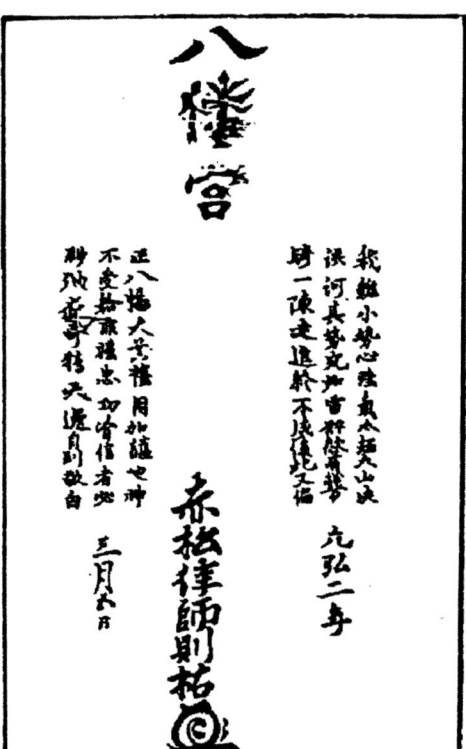

Heraldry as personal display

Until about the middle of the 15th century the only way in which a mon or a flag design was used to identify a particular warrior or his followers was by display on a hata jirushi banner carried by an attendant, on the maku, and on the large wooden shields used by footsoldiers on a battlefield. From the Nanbokucho Wars onwards the nature of warfare itself began to change, with increasing use being made of large numbers of footsoldiers

Akamatsu Norisuke, the son of Norimura, used this banner with an invocation of Hachiman.

RIGHT **The *shisume* (four eyes) *mon* of Amako Tsunehisa (1458–1541) and his descendants – a family wiped out by Mori Motonari and his sons in a long series of wars.**

FAR RIGHT **The abstract design of *mon* used by Miyoshi Chokei Norinaga (1523–64), one of the most prominent of the Sengoku daimyo during the first half of the 16th century.**

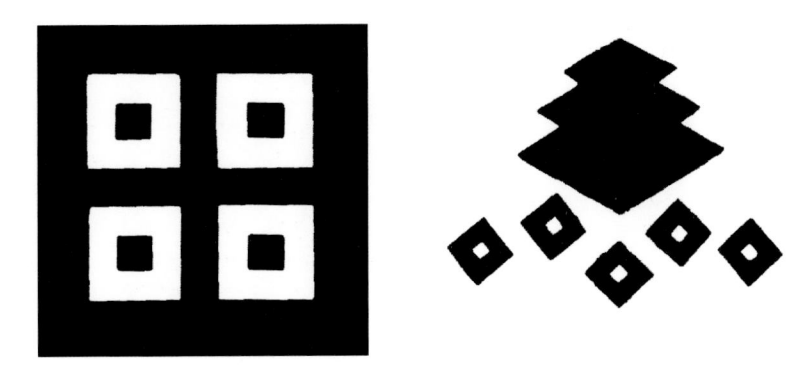

This is one of the two banners used by Ouchi Yoshitaka (1507–51), one of the first of the Sengoku daimyo. It bears the names of the deities Myomi Dai Bosatsu, Hachiman Dai Bosatsu, Amaterasu Dai Jingu, Sumiyoshi Dai Myojin and Kasuga Dai Myojin; but it also displays below the inscriptions the Ouchi *mon*, a form of heraldic identification that was to become almost universal during the 16th century. For Yoshitaka's other banner see Plate B5.

and a commensurate decline in the role of the individual mounted warrior. Many of these footsoldiers were used as archers, shooting massed volleys of arrows as the Mongol invaders had done. Mounted samurai then began to adopt the spear rather than the bow so that they could take the fight to the footsoldiers at close quarters.

If samurai had to fight in conditions where they were less likely to be recognised by means of an attendant's banner then some additional form of identification was needed. This was at first provided by the wearing of little flags, like tiny hata jirushi, attached to the shoulder plates of their armour or to a ring on the back of the helmet. The first type was called a *sode jirushi* ('shoulder plate insignia'), the second a *kasa jirushi* ('helmet insignia'); kasa jirushi could also be flown from the crest on the front of a samurai's helmet.

The use of these little flags is the first example in Japan of heraldic identification being worn on the person by an individual warrior, rather than having it proclaimed by the flag carried by his attendant. There is also evidence that similar flags were worn by a samurai's followers, thus making these small flags the first form of 'military uniform' seen in Japan. The use of the hata jirushi continued, but it was confined to the function that it had previously served under the Taira and Minamoto – as a flag representing the samurai's contingent of followers, or even a whole army, rather than an individual fighting man.

15TH–16TH CENTURY HERALDRY: THE RISE OF THE *DAIMYO*

The pivotal event of the 15th century in Japan was the Onin War (1467–77). This terrible struggle laid Kyoto waste and gravely weakened the power of the Ashikaga family. Although the institution of the shogunate was to last for another century, the shoguns gradually began to resemble puppets, and were cynically manipulated by the new breed of regional rulers who called themselves *daimyo* ('great names'). Daimyo were warlords who achieved power, prestige and possessions by the exercise of their military skills as much as by political manoeuvres or appeals to an ancient lineage. Very soon Japan had split into a number of warring territories; the resulting contention between ever-shifting alliances of the great regional families, prosecuted by open warfare,

(1) The *ginko* (maidenhair) device of Tokugawa Hirotada (1526–49), father of Ieyasu, who used it before adopting the *aoi* motif. Hirotada fought against Oda Nobuhide, Nobunaga's father.

(2) The sun design of the great Kyushu daimyo Ryuzoji Takanobu (1530–84) – a Christian, but noted for his cruelty, he was killed by the Shimazu at the battle of Okita Nawate.

(3) The *kashi* (oak leaves) *mon* of Yamanouchi Kazutoyo (1546–1605), who fought at Sekigahara in 1600.

(4) The 'sliced melon' device of Takigawa Kazumasu, who opposed Hideyoshi following the death of Oda Nobunaga and was defeated at Kameyama in 1583.

conspiracy and treachery, lasted for generations. This bloody century is known as the *Sengoku Jidai*, the 'Warring States Period', by analogy with the term for a similar time in ancient Chinese history.

Some daimyo were the heirs of ancient families, but they were nonetheless forced to change with the times. Ouchi Yoshitaka (1507–51) from Yamaguchi, who ruled over the western half of the Inland Sea for many years before being spectacularly supplanted by the Sue (and later avenged by the Mori), provides an early example of a Sengoku daimyo with a fine pedigree, and this was reflected in his heraldry. The Ouchi flag (see Plate B5) displays the ancient family mon accompanied by a religious invocation in the form of a list of names of the kami whose benevolence is being invoked. One version has 'Myomi Dai Bosatsu, Hachiman Dai Bosatsu, Amaterasu Dai Jingu, Sumiyoshi Dai Myojin, Kasuga Dai Myojin'; a later version has 'Myomi Dai Bosatsu, Hachiman Dai Bosatsu, Tenjin' (the deified Sugawara Michizane, who became the god of learning).

The Ouchi were eventually supplanted by Sue Harukata, who was an opportunist from among their own army. Such men were the other sort of daimyo who appeared on the scene, the ones who rose from obscurity by force and intrigue. The outstanding example of such adventurers is provided by a certain Ise Shinkuro, who overthrew his own daimyo and set himself up in the lord's territories in a classic illustration of the process of *gekokujo* ('the low overcome the high'). Ise's triumph was recorded in his heraldry: he adopted the surname of the famous and now extinct Hojo of Kamakura, simply because it had been the name of an illustrious samurai family in the past. Not only did Ise Shinkuro appropriate the Hojo name, he also helped himself to their mon of three triangles (see Plate C1); and as the emblem of the usurper now known as Hojo Soun it was to become better known in the wars of the Sengoku than it had ever been during the Nanbokucho Wars of the 14th century.

At the start of the Sengoku the actual means of displaying samurai heraldry was very similar to what had gone before, consisting of hata jirushi banners and the small personal flags attached to armour or helmets. Samurai warfare itself also changed very little, and still depended upon bow and sword, the only real innovation being the introduction of straight spears rather than the curved-blade *naginata* pole-arm for any footsoldiers who were not being used as archers.

One major development that occurred from the time of the Onin War onwards was that many daimyo took the opportunity presented by the lawlessness of the times to swell the ranks of their armies by the recruitment of irregular infantry. These newcomers were referred to as *ashigaru* ('light feet'), and were very useful if properly disciplined and led – otherwise they would run riot or desert. There were so many ashigaru available for hire that a warlord could easily increase the size of his army tenfold; but once the campaign was over these men would disappear as casually as they had arrived, and when the next campaign began they might well be found fighting for another family.

As time went by the more astute daimyo realised that these rough-and-ready infantry should be given the same continuity of employment as enjoyed by the daimyo's own long-standing retainers, who had worked his family's fields and fought their battles for generations. The result was that as armies increased in size, the training offered and the loyalty given in return also increased. Heraldic identification thus became even more important. As casually enlisted footsoldiers were not issued with armour until the mid-16th century, but had to supply their own, an early

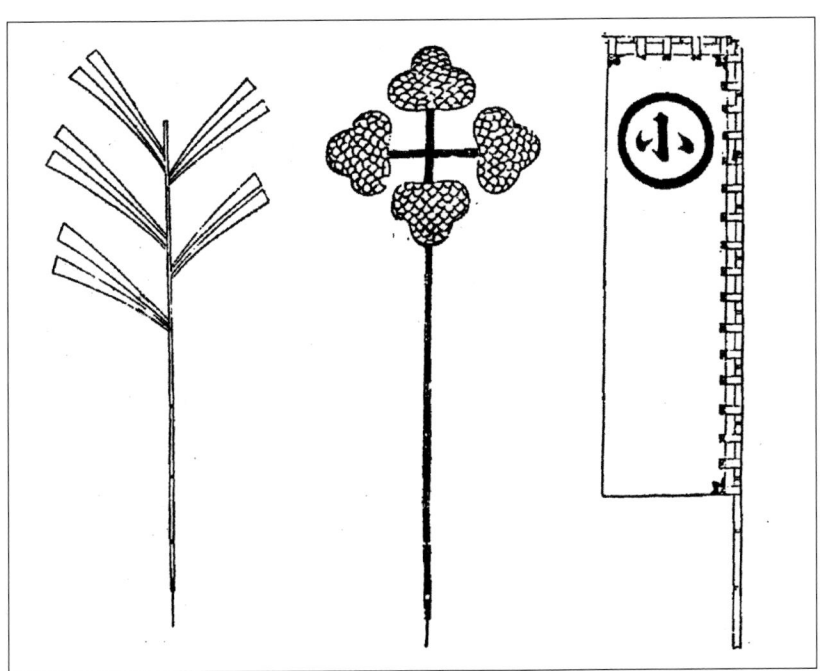

Koide Yoshichika fought at Osaka (1615) for the Tokugawa. His *nobori*, right, uses the first character of his name, *ko*, in a ring in black on white. The standard, centre, is an elaborate gold cross, while the *sashimono* worn by his samurai, left, is a tree of gold flags.

The most common form of *sashimono* consisted of a small flag. This illustration (after Sasama) shows how the shaft slotted into a holder fastened to the back of a samurai's armour using a hinged upper bracket and rigid lower one. For extra security in battle a cord was looped around the shaft and taken forward under the armpits; the ends were then tied to the hinged rings on the front of the breastplate.

ashigaru's individual identification was probably limited to the small shoulder or helmet flags noted above, with the army being organised overall under a daimyo's own banner. Eventually simple *okashi gusoku* ('loan armours') were issued to the ashigaru; these are described below.

Heraldry and army organisation

With the development of disciplined units of long service ashigaru, armies became not only larger but much better organised. A tremendous boost was given to this trend by the introduction of firearms from Europe after 1543. From about 1550 onwards, rather than forming the old amorphous mass of spearmen and archers, ashigaru began to be organised in specifically armed units with either the matchlock, the bow or the long spear. This increase in systematic deployment and control was reflected in the heraldry now seen on battlefields. In fact, far from merely echoing developments in army organisation, heraldry became one of the main means by which it was carried out, along with drum signals and calls from conch shell trumpets. There was a more widespread and systematic use of coloured flags bearing geometrical shapes and divisions instead of, or together with, the mon. Different coloured flags were also now used by sub-units, so from this time onwards it became necessary to identify more than just a mon when 'reading' the flags carried by an army.

So important was the heraldic dimension than from the 1550s onwards large numbers of footsoldiers were employed simply to carry flags to aid an army's dispositions on the battlefield. This is shown by the muster roll of Uesugi Kenshin in 1575: out of a grand total of 6,871 men, of which 6,200 are footsoldiers, no less than 402 are employed in carrying flags – more than the number of arquebusiers. So crucial was the correct organisation and carrying of flags in a daimyo's army that a senior samurai would be given the post of *hata bugyo* ('flag commissioner'), with the task of ensuring that all flags, standards and other devices were in place and displayed according to the agreed regulations.

A few examples of *sashimono* with samurai's names on them. These men are fighting during the siege of Kanegasaki Castle in 1570.

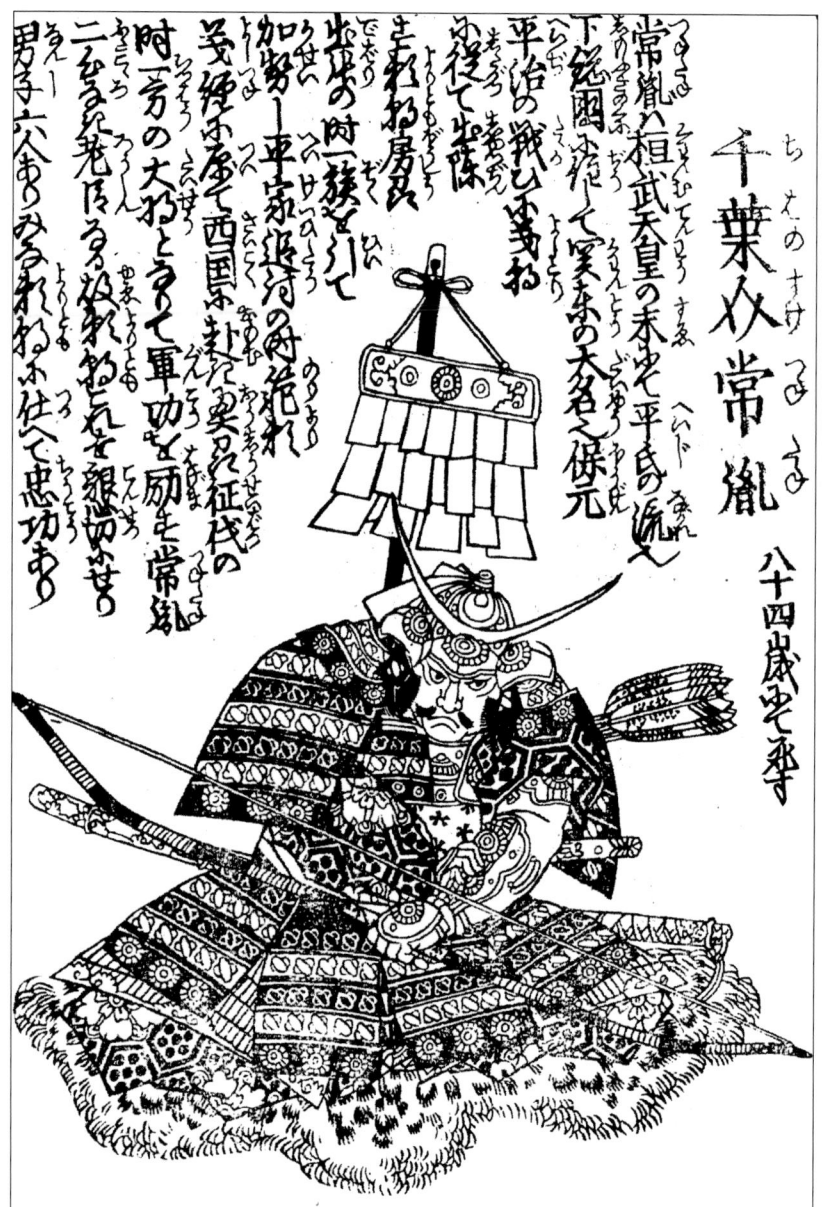

There were many variations of *sashimono*, ranging from flags to three-dimensional objects. Here we see one in the form of a small *hata jirushi*, slung from a crossbar.

The *nobori*

The flags on which the mon were displayed therefore became more numerous, and new types of flags were also developed. In addition to the familiar hata jirushi banners we begin to see considerable numbers of *nobori* – tall, narrow flags of which two edges were attached by loops to both a vertical shaft and a short crosspiece at the top, so that the flag could not be obscured in a high wind by getting wrapped around the pole. Nobori were carried with the shaft supported either in a leather socket strapped at the bearer's waist or, in the case of larger versions, in a specially strengthened holder on his back (as described below for the *sashimono*). Nobori were most commonly used for identifying the subdivisions within an army, or simply to provide a grand heraldic display of scores of identical banners. On a painted screen depicting the Takeda army drawn up ready for the battle of Kawanakajima in 1561 there are two lines of ashigaru who are holding only nobori, and appear to have no other function.

The *sashimono*

While the nobori and the older style of hata jirushi now provided identification, and allowed easy visual organisation, of largely ashigaru units, another new type of flag was introduced to give units of samurai spearmen and swordsmen a more prominent visual signature. This was the *sashimono*, the most important heraldic innovation introduced during the Sengoku period. Most sashimono consisted of a small flag similar to a nobori which was held rigid by two poles threaded through loops on the edges of the cloth, and flown from the back of a samurai's suit of armour. In function the sashimono thus replaced the small kasa jirushi and sode jirushi, and their larger size made identification easier (which was particularly necessary because of the greater similarity in the design of the armours that were now being worn throughout Japan).

The sashimono was slotted into a lacquered wooden carrying socket securely fastened to the back of the cuirass, and was held in place by the

A fine example of a back-flag being used to convey a message and an intention rather than mere identification – in this case the name of the samurai (Nakagawa) and his intended victim (Wada), whom he is seeking out to kill in battle.

The painted screen of the battle of Shizugatake (1583) which is displayed in Osaka Castle Museum is an immensely rich source of information about samurai heraldry. This is the right-hand panel showing the final stages of the action. On the hill to the left rear is the army of Shibata Katsuie, who is seen waving a red fan. The pursuit by Hideyoshi continues over to the far right, where the red *nobori* with three black *mon* of Sakuma Morimasa appear quite prominently. In the valleys below various units of Hideyoshi's army charge uphill.

two ends of a cord passed under the armpits and tied to the metal rings on the samurai's breastplate. Sashimono could be a hindrance during close quarter fighting, and a painted screen in Hikone Castle which illustrates the Summer Campaign of Osaka in 1615 shows a samurai's attendant holding his master's sashimono while he engages in personal combat.

The sashimono usually bore the mon of the warrior's daimyo, thus giving the whole family army something of a uniform appearance, the units being distinguished from one another by the field colours of the flags. The five 'lucky colours' of red, blue, yellow, black and white were the usual choices; e.g., the three-triangle mon appeared on the sashimono of the 'five colour regiments' of the Hojo. In the case of the Ii family of Hikone, however, the ground colour of red was itself a more important distinguishing feature than was the Ii mon; consequently certain Ii samurai were allowed to display their own mon, or to have their names written in long flowing golden characters. The sashimono could also provide unit identification by added bands in black or other colours, as on the old wooden shields; or even by a character from the Japanese alphabet – a method used by the Hojo for distinguishing another section of their army.

Sometimes the demands of army uniformity would be overruled, and a samurai who wished to draw attention to himself for reasons of personal revenge, or to fulfil a vow by fighting in what might be his last battle, would be allowed to write his name on a plain white flag and use it as his sashimono – see the example on page 23.

There were numerous examples of sashimono taking the form of three-dimensional objects rather than flags. Plumes of feathers, wooden

A modern reproduction of Toyotomi Hideyoshi's 'thousand gourd' standard; it is gold with a hanging fringe of red strips. He is supposed to have added one gourd for every victory he gained after Inabayama in 1564, when his men waved their water gourds on their spears as a signal of victory.

gourds and golden fans are noted on painted screens; and a certain Tsuji Hikobei, a retainer of Takeda Shingen, used as his personal sashimono a golden *ema*, a larger version of the wooden prayer boards found in Shinto shrines. Tsuji's ema was gold, and bore an inscription supposedly written by Takeda Shingen himself: 'He who fights skilfully will not die'. Other families used golden discs, or *gohei* (the Shinto prayer wand), suspended from the same poles that were otherwise used for the sashimono flags.

Battlefield accounts in the war chronicles frequently mention sashimono flags and their heraldic designs. A good example comes from a description of the capture of P'yongyang in 1592 during Hideyoshi's Korean campaign, found in the *Kuroda Kafu*, the chronicle of the Kuroda family:

'At their head were Goto Mototsugu, Yoshida Rokurodaifu and Toda Heiza'emon who crashed into the enemy. Mototsugu wore a helmet which had a crest of two golden irises, and couched his long-shafted spear from his horse. Rokurodaifu wore a sashimono of a crane within a circle and a crest of a polar bear on a shaft. The two men vied with each other over the lead position, then plunged into the midst of the enemy, wielding their spears in every direction.'

It was normally only very high-ranking samurai such as Tsuji or Goto Mototsugu who could be recognised by their own mon on a personal design of sashimono. Daimyo would occasionally wear a sashimono if they were expecting to be in the thick of battle, and would abandon their customary jinbaori surcoat in favour of a back-flag, since both could not be worn at the same time. For example, Matsuura Shigenobu, the daimyo of Hirado, had a personal sashimono showing a gold disc on a black field. Otherwise a daimyo was known by his standards – another 16th-century innovation.

'GREAT' AND 'LESSER' STANDARDS

With so many flags now appearing on a battlefield an important samurai's personal identification had to be more prominent than ever if

(1) The *mon* of Uesugi Kenshin (1530–78), the great warlord of Echigo, showing the 'birds in bamboo' motif.

(2) The *take ni suzume* (birds in bamboo) *mon* of Date Masamune (1566–1636), the 'One-Eyed Dragon' from Tohoku; this famous northern daimyo was one of the last to submit to Hideyoshi after his 1589–90 Odawara campaign. Compare this with the Uesugi version.

(3) The 'Bishamon-ten' flag of Uesugi Kenshin, showing the first character of the name of the *kami* or Shinto deity to whom Kenshin was particularly attached.

1

2

3

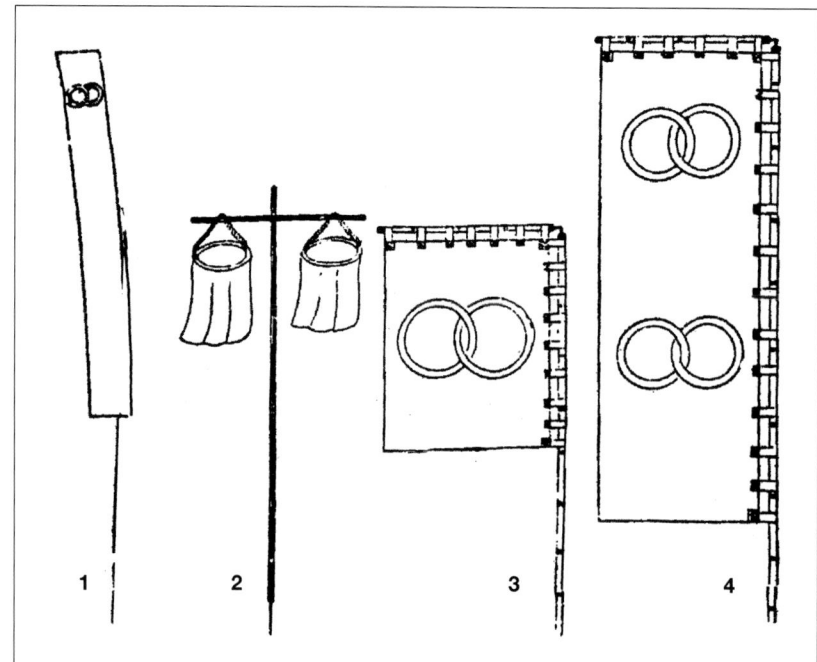

LEFT **Wakizaka Yasuharu (1554–1626) was one of the 'Seven Spears' of Shizugatake in 1583, and went on to fight in Korea, where he was defeated at the naval battle of Hansando (1592). All his flags used the white** *mon* **of conjoined rings on a red ground.**
(1) Samurai *sashimono*
(2) lesser standard – two white paper lanterns
(3) great standard
(4) *nobori*.

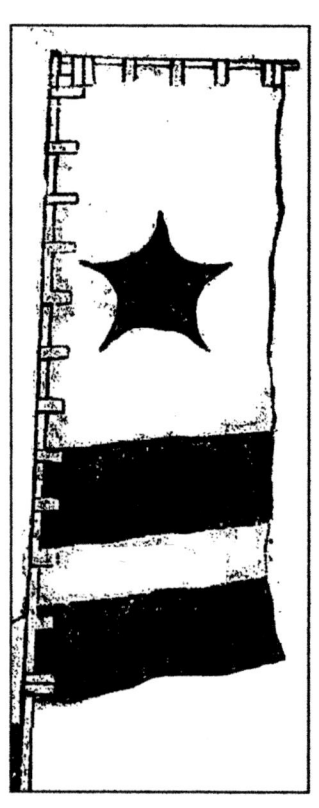

ABOVE **The standard of Yamanaka Shika, the pillar of loyalty of the Amako family who died in 1578. In** *nobori* **form, it bears a star design.**

ABOVE **The two flags bearing the character** *tai*, **'great', used as standards by Takeda Katsuyori (1546–82), who was defeated by Oda Nobunaga and Tokugawa Ieyasu at the battle of Nagashino in 1575.**

LEFT **An interesting array of different heraldic devices: a rectangular flag bearing the** *mon* **of Ikeda; a three-dimensional umbrella standard; and a large** *fukinuki* **or hollow streamer. This latter appeared in the battlefield display of many different daimyo.**

it was to stand out at a glance. This was particularly necessary for a daimyo; and the most spectacular sight on a samurai battlefield were now the warlords' large battle standards. By the early 17th century most daimyo would have two: an *o uma jirushi* ('great standard') and a *ko uma jirushi* ('lesser standard'). These were sometimes large rectangular or elongated nobori-type flags, but were more usually large three-dimensional objects, often made from light wood, in the shape of bells, umbrellas, gongs or other symbolic objects. The overall display within a daimyo's headquarters would also include other flags and banners; and the sight in contemporary illustrations of a collection of uma jirushi set up within a daimyo's maku is very reminiscent of the *carrocio* of contemporary Italy.

For example, in addition to his uma jirushi of a dark blue flag with a sun motif, Uesugi Kenshin cherished two other banners. One bore the character *bi*, the first ideograph in the name of

A standard in the form of a *gohei*, the ceremonial 'wand' of a Shinto priest – a popular choice of religious object for use in heraldry. It is made of gold-lacquered paper, and may be compared with the standard of Shibata Katsuie (1530–83) – see right.

RIGHT In this print the hero Menju Ietora fights to win back Shibata Katsuie's golden *gohei* standard, which had been captured by the enemy.

Bishamon-ten, a deity to whom Kenshin was devoted; and the other bore the Chinese character for 'dragon'. This flag was flown to send the Uesugi samurai off in a charge against their enemies. Kenshin's great rival Takeda Shingen similarly cherished two nobori flags which were associated with the kami Suwa Myojin. He also used one of the earliest examples in Japanese history of the *hi no maru*, the sun disc banner that was to become Japan's national flag.

Many 'great standards' became very famous, and were instantly recognisable to friend and foe alike. Takeda Shingen used as his o uma jirushi a large blue flag bearing the quotation from the ancient Chinese strategist Sun Tzu: 'Swift as the wind, deadly as fire, silent as the forest and steady as the mountain', which was Shingen's motto. Tokugawa Ieyasu was known by a huge golden fan giving the effect of the rays of the rising sun (see Plate H1); while Oda Nobunaga used a large red umbrella (see Plate D1). His son Oda Nobuo's golden fan was described as 'rising like a sun' when he marched into Ise province. Toyotomi Hideyoshi, the 'Napoleon of Japan', used a gold-lacquered gourd shape as his standard (see Plate E1), in recognition of his bravery at the siege of Inabayama in 1564 – he had led a detachment of men up a narrow mountain path, and they signalled to Oda Nobunaga's troops waiting below by waving their water gourds from the end of their spears. The final form of Hideyoshi's standard was the striking 'thousand gourd standard', last carried in battle by his doomed son Hideyori at Osaka in 1615. This was a cluster of many (though by no means a thousand!) golden gourds, each representing one of his victories.

Other daimyo chose objects that had personal significance to them. Konishi Yukinaga, who led the invasion of Korea in 1592, was the son of a medicine dealer from Sakai, so for his uma jirushi he displayed a huge white paper bag as used by pharmacists in Japan, with a large red sun painted on it. Kato Kiyomasa, who was a devout follower of the Nichiren sect of Buddhism, used a centuries-old white hata jirushi with the inscription 'Namu Myoho Renge Kyo' – 'Hail to the Lotus of the Divine Law' – said to have been written on it by Nichiren himself (see Plate G1). Tsugaru Tamenobu from the far north of Japan fought under a gigantic *shakujo*, the metal 'rattle' carried by the itinerant *yamabushi* sectarians to frighten away dangerous animals while on their mountain pilgrimages (see Plate I1). The Omura of the Nagasaki area used a huge golden bell; while Ankokuji Ekei, who had once been a Buddhist monk, sported a large golden lantern.

The larger varieties of uma jirushi were very heavy and often unwieldy to carry, but there does not appear to have been any equivalent Japanese device resembling the cart that provided the centrepiece of the Italian carrocio. Instead, several surviving painted screens show standard-bearers with their master's uma jirushi strapped

The samurai *sashimono*, great standard and *nobori* used by three generations of the Omura family of Nagasaki: Sumitada (1532–87), the first Christian daimyo; Yoshiaki (1568–1615), who fought in Korea; and Suminobu, who so thoroughly rejected Christianity that he fought against the Christian rebels at Shimabara in 1638. The flower device is picked out in black outline on white, while the standard is a huge three-dimensional golden bell – cf that used by Mukai Tadakatsu, illustrated as Plate H15.

to their backs in specially strengthened sashimono holders, as noted above for nobori. The bearer would steady the contraption by using two long ropes; and in the case of the largest types extra bracing ropes would be held by two other footsoldiers. The *fukinuki*, a type of huge streamer rather like the device in the form of a carp used at the Boys' Festival, was particularly difficult to manage in a high wind because it filled up like a windsock.

Helmets

The final form of heraldic identification to be developed during the Sengoku period concerned the samurais' helmets. The tendency towards uniformity apparent in the practical and sombre 'battle' armours worn by samurai at this time extended to the highest ranks of warriors. Wishing to make themselves noticeable above the common herd, senior samurai frequently enhanced a plain body armour by embellishing their helmets with many weird and wonderful designs of buffalo horns, peacock feather plumes, theatrical masks, conch shells, etc. When Admiral Yi of the Korean Navy won a battle against the Japanese in 1592 a collection of extraordinary helmets was among the booty taken, and Yi described them in tones of wonder in his report to his king.

Huge wooden horns were a popular embellishment, and appeared on helmets owned by Tokugawa Ieyasu, Kuroda Nagamasa and Yamamoto Kansuke, making their wearers look like the Vikings of cherished but misguided mythology. Stretched versions of courtier's caps or catfish tails were also popular. Kato Kiyomasa owned two varieties: one was silver with a sun disc on each side, and the other was black with Kato's 'snake eye' mon in gold. Maeda Toshiie had a tall golden helmet, and his son wore a similar silver one. Uesugi Kenshin wore a helmet with three masks from a Noh play displayed around the bowl. In every case the increase in size was obtained by moulding papier maché on to a light wooden framework above a simple iron helmet bowl.

A daimyo's helmet was often carried to the battlefield on the end of a spear, and thus acted as a further variation on the theme of a personal standard; but spectacular helmets were not always the close preserve of generals. One popular style added horsehair to the top of the bowl. This 'wig' was shaped and tied in a pigtail to look like a human hairstyle – a feature shown to best effect when combined with a mustachioed face mask and the curious beaten breastplates designed to look like a muscled human torso.

The helmet was also another place where mon would often appear. This was done by use of the *maedate* or crest, a badge made of thin metal, with the mon design either lacquered on to it or cut out in two dimensions. Mon were also often applied to the front of the *fukigayeshi*, the 'turnbacks' on each side of a helmet designed to catch downwards sword strokes.

The silver helmet of Kato Kiyomasa (1562–1611), in the elongated form of a courtier's cap and with a red sun disc on the sides, is an excellent example of the elaborate styles favoured by high-ranking samurai in the later Sengoku period. This was done largely to make them look particularly distinguished in battle, but as the helmet was often carried on a pole by an attendant footsoldier it could also function as a heraldic device.

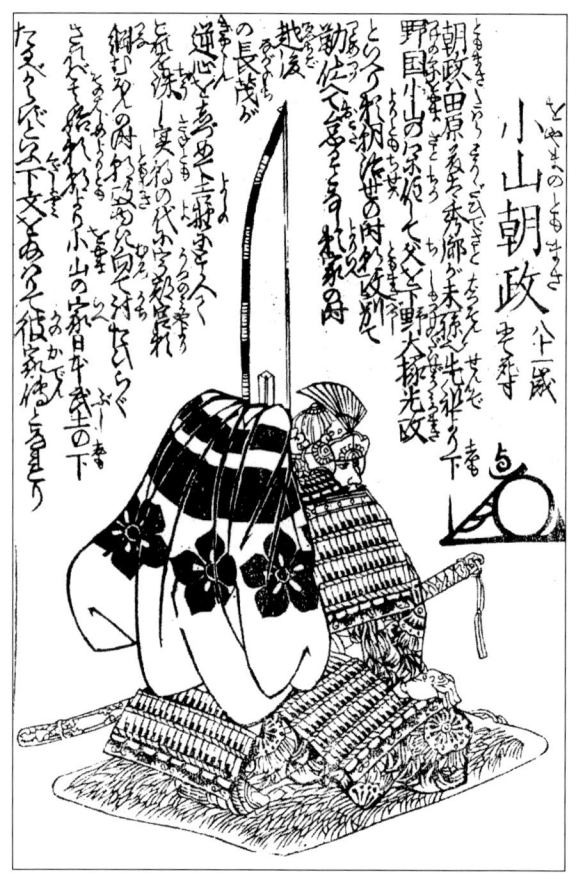

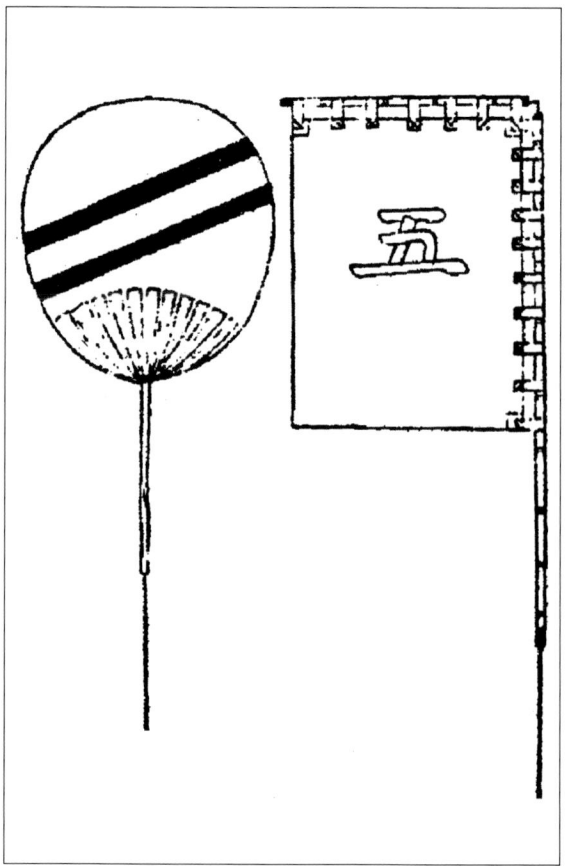

ABOVE **A warrior wearing a** *horo*, **a cloak stretched over a light bamboo framework which would fill with air as the man rode along.** *Horo* **were a popular means of identifying a distinguished warrior acting as a bodyguard or a messenger.**

ABOVE RIGHT **Instead of the** *horo*, **Tokugawa Ieyasu's messengers wore a** *sashimono* **(right) bearing the character** *go* **(five), referring to the god Fudo. The colours are variously illustrated as white on light blue or black on white. His personal bodyguard wore this** *sashimono* **(left) – not a flag, but a golden fan with black stripes attached to a pole.**

Elite samurai and the *horo*

Apart from a daimyo's collection of standards, the most gallant sight on a battlefield was the colourful appearance of the elite warriors who wore *horo*. This was another item of equipment that had a heraldic function, and consisted of a curious form of cloak stretched over a basketwork frame. The horo filled up with air as the horseman rode along, and was a popular means of identifying a daimyo's personal mounted samurai bodyguard. There is a reference in the *Hosokawa Yusai Oboegaki* to a particular use of both horo and sashimono: 'When taking the head of a horo(-wearing) warrior, wrap it in the silk of the horo, and in the case of an ordinary warrior, wrap it in the silk of the sashimono' – a passage that confirms the impression that a man who wore a horo was something special. Oda Nobunaga had two elite units distinguished by red and black horo, while Toyotomi Hideyoshi's bodyguard wore gold-coloured horo.

The elite *tsukai ban* (messenger corps), who acted as aides-de-camp on the battlefield and therefore had to be instantly recognisable, often wore either brightly coloured horo or an extra-large sashimono, or even both together. The horo might bear the daimyo's mon on a brightly coloured background, or would be parti-coloured in a very striking way. Instead of a horo Takeda Shingen's messengers wore a sashimono bearing the apt device of a busy centipede. Tokugawa Ieyasu's tsukai wore a sashimono with the character *go* (the figure '5') – a mystical number associated with the god Fudo.

THE STANDARD BEARER IN BATTLE

Because of their high visibility and their vital functions the brave standard-bearers attracted the heaviest fire on battlefields, and their roles are often mentioned in the war chronicles. The Akamatsu family provides one of the earliest examples of a written account of bravery associated with a samurai lord's standard-bearers, who were then known as *hata sashi* ('flag carriers'). The following anecdote from the *Meitoku ki* refers to a battle at the end of the Nanbokucho conflicts in the late 1380s between Akamatsu Yoshinori (1358–1427), son of Norisuke, and Yamana Ujikiyo (1345–92):

'The flag-bearer of Kazusa no suke [Akamatsu Yoshinori] plunged into the midst of the multitude, and, taking hold of the flag by its shaft, cut around him vigorously; he then galloped up to the flag-bearer of Oshu [Yamana Ujikiyo], inevitably clashing violently with him. They stabbed at each other, and both men died heroically.'

In other accounts from the Nanbokucho Wars we read of standard-bearers being hit in the face and wounded by rocks thrown from castle walls, and the sheer number of such references to close quarter action confirms that flag-carriers were always in the thick of the fighting.

On at least two occasions in Japanese history a standard-bearer and his flag played a part in attaining the supreme samurai glory: that of being first into battle. In the case of an assault on a defended castle, such an exploit was referred to as *ichiban nori* ('first to climb in'), and could be claimed for oneself even if only one's flag had actually made it into the castle!

An excellent illustration of a samurai attended by a foot-soldier carrying his standard. His name is given as Kamikoda, and he is pictured at the battle of Shizugatake in 1583.

神田半左門通清

The *take* (bamboo) *mon* of Yamana Ujikiyo (1345–92), which would have appeared on the *hata jirushi* standard carried by his brave flag-bearer in the battle against the Akamatsu described in the text.

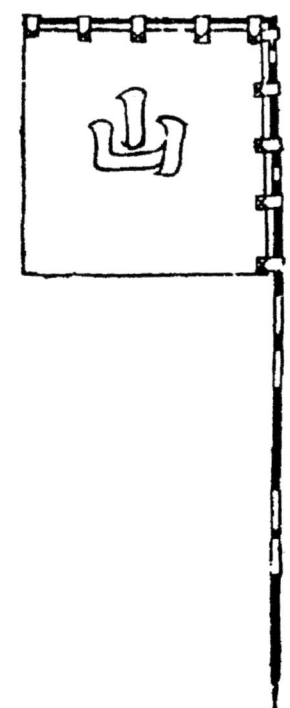

The great standard of Aoyama Tadanari (1551–1613), who served Tokugawa Ieyasu and was raised to the status of daimyo in 1601. His flag shows a nice example of a heraldic pun (albeit in reverse), as his name literally means 'blue mountain' and the charge is a white character *yama* (mountain) on a blue ground. Similar smaller flags were used as *sashimono* by his samurai and messengers.

The first example occurred during the second siege of Chinju in Korea in 1593. So desperate was the rivalry that squabbles broke out between allied contingents, with samurai pushing each other off the scaling ladders for the glory of leading the assault. Kato Kiyomasa's unit were distinguished by the long white hata jirushi which bore the slogan 'Namu Myoho Renge Kyo'. Ten brave warriors took it in turns to carry this precious Nichiren flag, and at Chinju the honour had fallen to Iida Kakubei. Seeing that Goto Mototsugu, a retainer of Kato's rival Kuroda Nagamasa, was likely to be the first to climb in, Iida threw the Nichiren flag over the wall to claim his place – see Plate G.

A similar incident of 'following the flag' occurred during the attack on Gifu in 1600. During the first assault Ikeda Terumasa had achieved the distinction of ichiban nori, to the fury of Fukushima Masanori, who was determined to have one of his men enter the castle first when the attack resumed. Ikeda was again ordered to lead the vanguard, but Fukushima charged in hot pursuit, and his men threw their flags over the wall just to make sure.

The hata bugyo ('commissioner for flags') could also play a part in rallying a retreating army, as on the occasion described in the following account from the *Shahon Heiyo Roku*:

'At the time of the Winter Campaign of Osaka, when Todo Izumi no kami's vanguard were routed, the Todo family's hata bugyo Kuki Shirobei hoisted three of the flags he was entrusted with, and stepped forward into the middle of the fleeing men. He set up the flags where they could be seen, and Shirobei knelt down, with the flags in front of the vanguard standing firm, and the soldiers who had been retreating were encouraged by this and came to a halt. Soon they turned back, dressed their ranks, and were again fighting men.'

Standard-bearers were often lowly ashigaru without surnames, but since they often had opportunities to show great bravery in guarding, or even regaining a master's uma jirushi their promotion prospects were greatly enhanced by service in this dangerous role. Similarly, for a samurai to take the head of a standard-bearer was a feat akin to decapitating a high-ranking enemy commander. Menju Ietora, the standard-bearer to Shibata Katsuie, risked his life to recapture Shibata's standard of a huge golden gohei when it was lost in battle (see page 27). During the Korean campaign in 1597 Shishido Bizen no kami lost his standard of a white swan to the Chinese army at Ulsan, and then lost his life trying to retrieve it. In another battle the Ukita standard-bearer suffered the disgrace of having the standard 'knocked out of his hands'.

Ashigaru heraldry

The lowly ashigaru weapon companies provided the most rigid example of rules of heraldic display. From about 1560 onwards ashigaru were being issued with 'loan armour', usually of the simple *okegawa-do* style. These cuirasses consisted of a number of horizontal iron plates riveted together to make a breastplate and a backplate. The resulting smooth, slightly ridged surfaces were lacquered, and would almost invariably have a common identifying motif – usually the daimyo's mon – stencilled on the breastplate. Some accounts state that the mon was also applied to the backplate, but the present author has never seen any examples of this.

(continued on page 43)

THE BATTLE OF DAN NO URA, 1185
See text commentaries for details

B

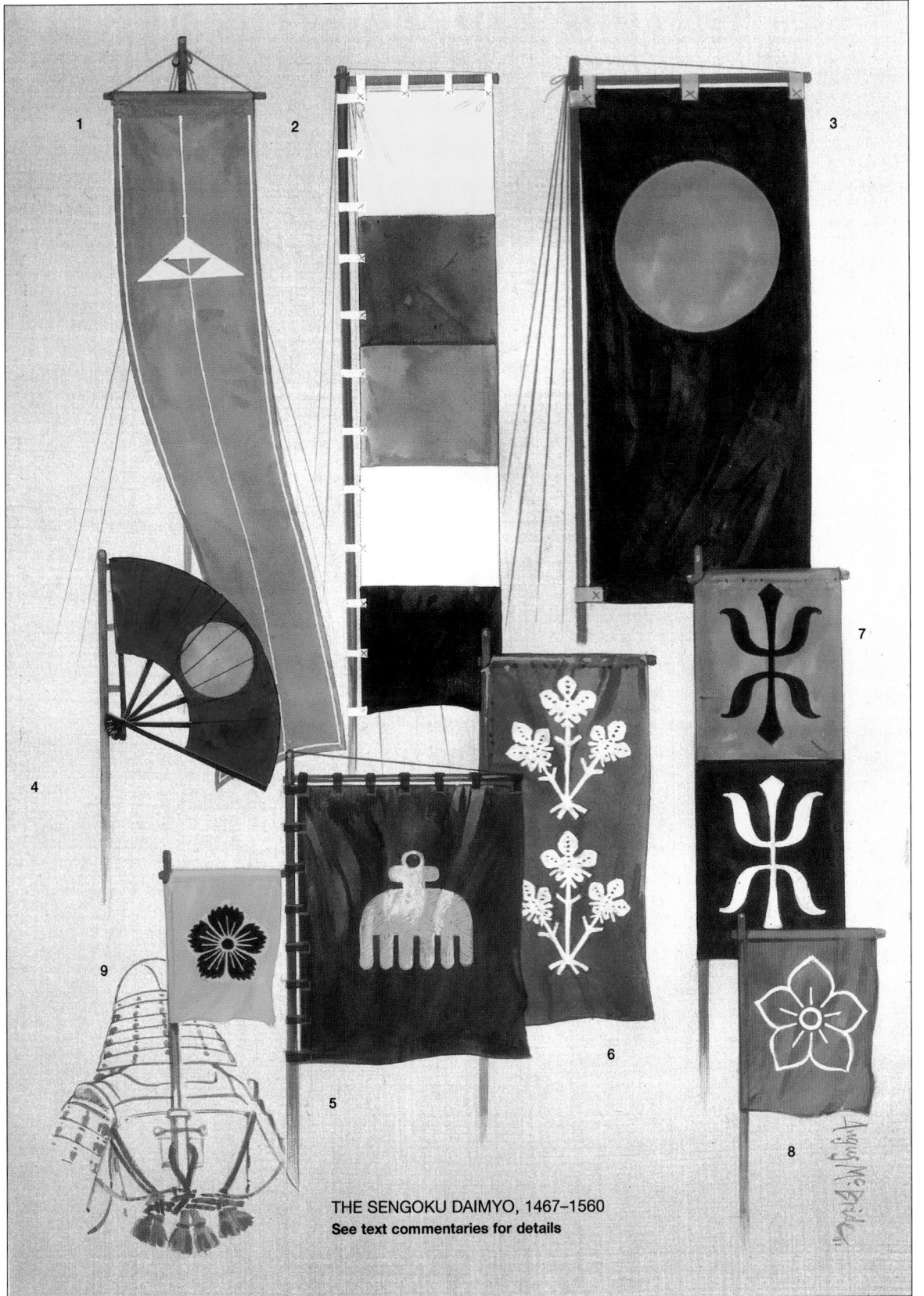

THE SENGOKU DAIMYO, 1467–1560
See text commentaries for details

C

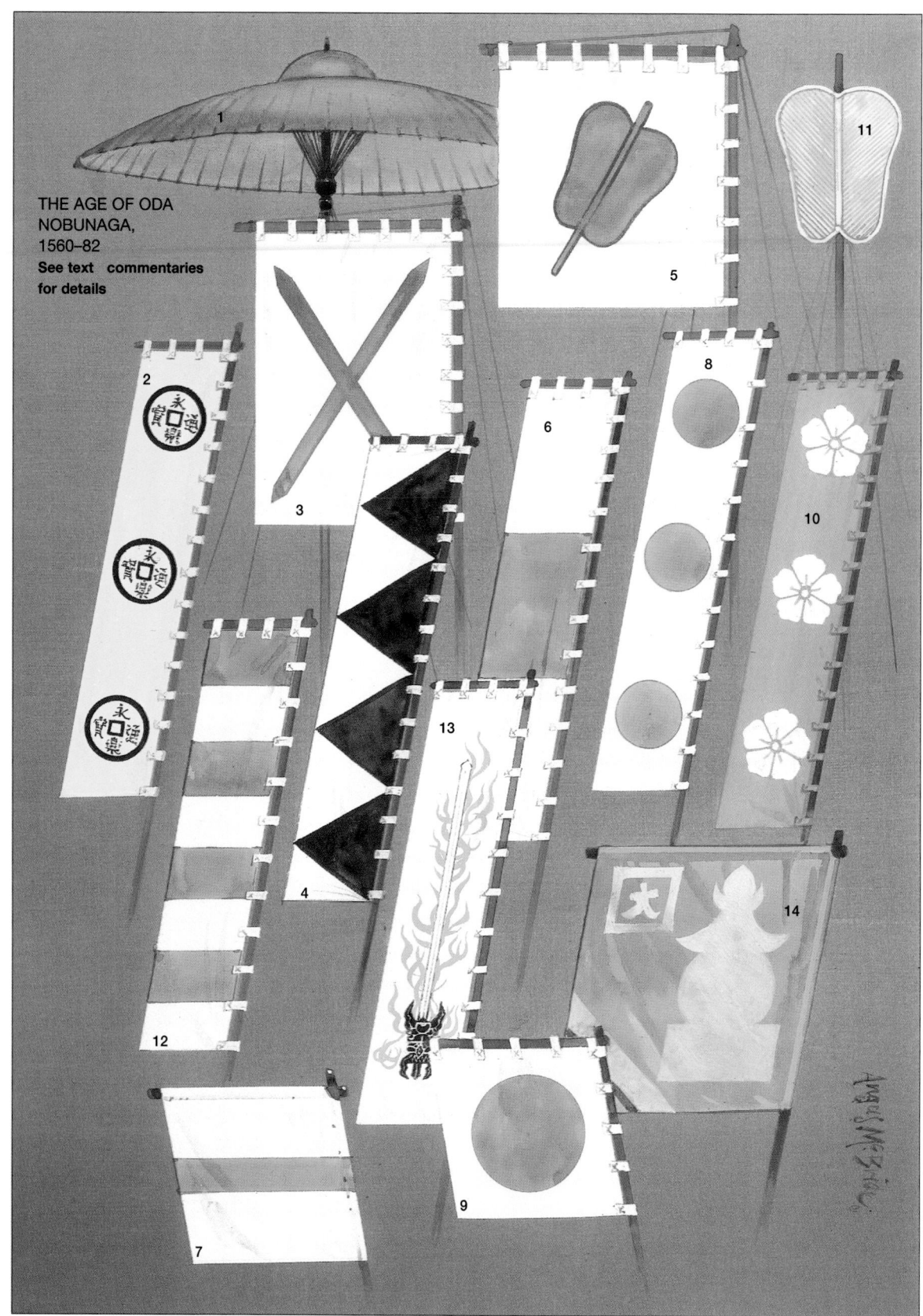

THE AGE OF ODA
NOBUNAGA,
1560–82
**See text   commentaries
for details**

D

THE AGE OF
TOYOTOMI HIDEYOSHI,
1582–98
See text commentaries
for details

E

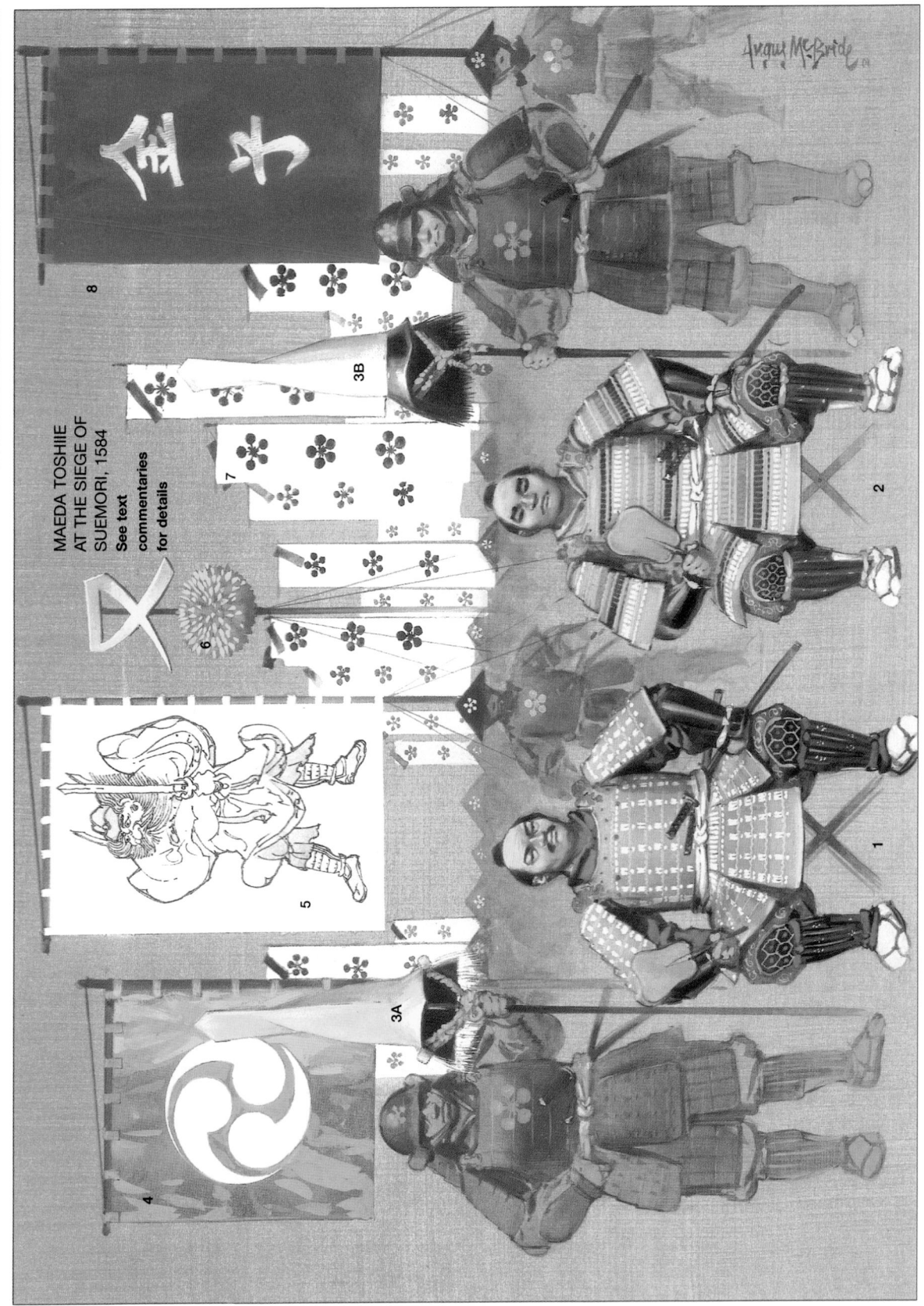

MAEDA TOSHIIE
AT THE SIEGE OF
SUEMORI, 1584
See text
commentaries
for details

KATO KIYOMASA
AT CHINJU, 1593
See text
commentaries
for details

1

2

3

Angus McBride

G

THE AGE OF TOKUGAWA
IEYASU, 1598–1615
See text commentaries for details

H

TSUGARU NOBUHIRA AT HIROSAKI CASTLE, 1610
See text commentaries for details

Louvar Seiao Sactissim Sacramento

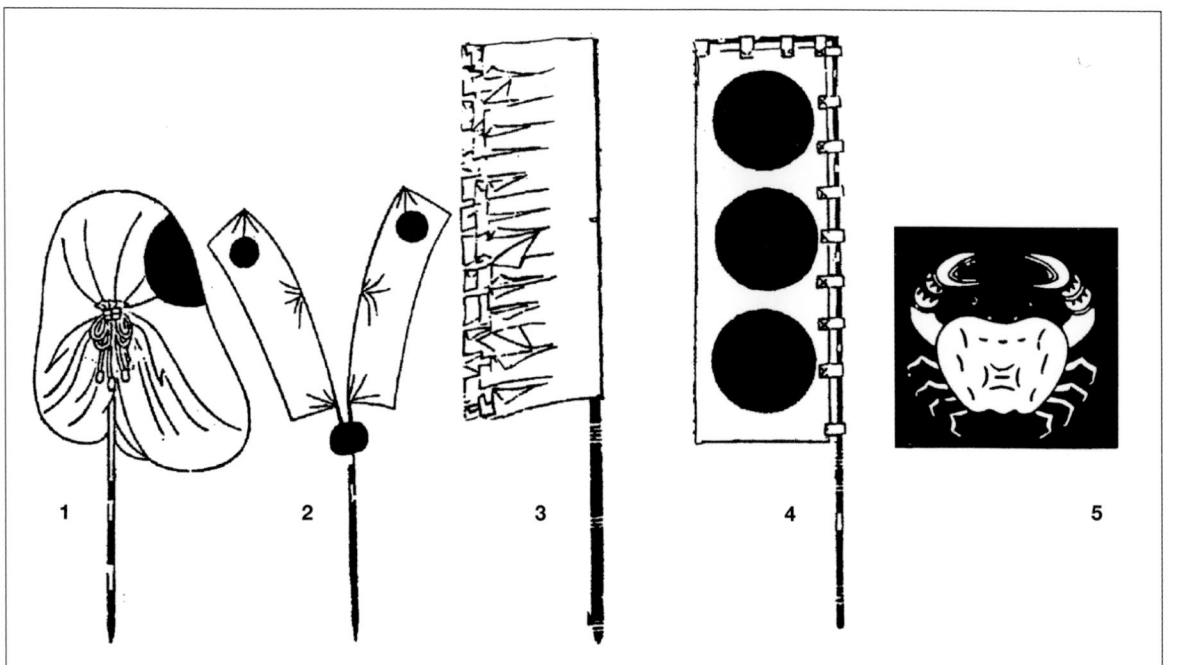

Terazawa Hirotaka (1563–1633) was governor of Nagasaki and took part in the Korean War, but his largest contribution to that campaign was to see his fief of Karatsu disappear under the huge supply base of Nagoya which was built for the invasion. His messengers' *horo* (1) is red with a black disc; his foot-soldiers wore two white flags with black discs (2); his standard is a double white flag with serrated edges (3); and his *nobori* is white with three black discs (4). His *mon* of a crab, *kani* (5) is unusual – animal motifs are seldom seen in Japanese heraldry.

In most cases these designs provide the clearest examples of mon being used for heraldic purposes, but there were some interesting alternatives. During the Korean campaign some of Kato Kiyomasa's men wore armour with the Nichiren slogan lacquered on their breastplates, a design proudly noted by the chronicler of the action in Manchuria in 1592.

Ashigaru sometimes, but not always, also wore two or three small flags as identifying sashimono on the back of their simple 'munition armours'. Some daimyo went to the lengths of dressing all their soldiers in armour of identical colour and design; the best-known example are the Ii family, with their famous 'red regiment'.

RELIGIOUS SYMBOLISM IN HERALDRY

Religious symbols and motifs were widely adopted by samurai families as their mon. The Sakakibara used the Buddhist 'wheel of the law' – *horin*, while the Hachisuka and the Tsugaru adopted the ancient Buddhist emblem of the swastika. The *tomoe* ('comma') design, the traditional shape of the sacred jewels of the emperors, also appears singly, in pairs or in threes on various mon such as that of Kobayakawa Takakage, the double tomoe forming the famous *yin-yang* shape. An ancestor of the Narita family is supposed to have eaten from a bowl of rice that was an offering at a shrine, and then went on to victory. He therefore adopted the mon of a rice bowl crossed by two chopsticks, a device that appears in a stylised form in the heraldry of several other samurai families, including that of Nitta Yoshisada who captured Kamakura in 1333. The Torii family used a mon of a *torii*, the Shinto shrine gateway, as a heraldic pun. Tachibana Muneshige's mon was a *mamori* (shrine amulet). In addition, the style of banners noted during the Gempei Wars that bore

The rice bowl with chopsticks design, used by several famous samurai including Hosokawa Katsumoto (1430–73), one of the protagonists in the Onin War, and Ashikaga Yoshiteru (1535–65), the 'swordsman shogun'.

In this illustration a group of samurai drawn up behind mantlet shields at the edge of a wood have disguised themselves as members of the fanatical Ikko-ikki sectarian movement. Part of the illusion is to fly banners bearing invocations of the Buddha.

invocations of Shinto deities, such as Hachiman Dai Bosatsu, remained just as popular during the Sengoku period three or four centuries later.

As noted earlier, several daimyo used Buddhist and Shinto names and motifs on their uma jirushi. Both Takeda Shingen and Uesugi Kenshin were ordained as Buddhist monks during their military careers, and reflected their faith through their heraldic displays. 'Layman' daimyo often did the same. Tokugawa Ieyasu proudly flew a hata jirushi bearing the slogan 'Renounce this filthy world and attain the Pure Land', an allusion to the Jodo sect of Buddhism to which he belonged. Maeda Toshiie had a flag bearing a picture of Shoki, the mythological queller of demons (see Plate F5); and there are many examples of the names of the Ise and Kasuga shrine deities being carried on flags.

At the other end of the social spectrum the armies of warrior monks from the Ikko-ikki, the Jodo Shinshu sect who opposed Oda Nobunaga during the 1570s, had their own very distinctive form of heraldry. Little use was made of family mon, as they drew most of their recruits from among the rebellious peasantry. Instead we see simple mon that have a religious connection, and a vast array of flags with Buddhist slogans. 'Namu Amida Bosatsu' was a frequently noted invocation upon Ikko-ikki banners; and one contingent used the device of a *sotoba*, the Buddhist cosmological 'pyramid' (see Plate D14). These fanatics held out at the fortress cathedral of Ishiyama Honganji for almost a decade in Japan's longest siege.

The Ikko-ikki armies and also their supporters (most notably the Mori family) made use of the defiant motto 'He who advances is sure of heaven, but he who retreats will go to hell'. During the battle of

A flag bearing the inscription 'Kasuga Dai Myojin', the *kami* of the Kasuga shrine in Nara – a popular choice for a religious motif.

RIGHT A *sashimono* with the name of Hachiman, the god of war, personally presented to Sanada Nobutada (the brother of Sanada Masayuki) by Takeda Shingen of Kai. The Hojo family of Odawara used a similar flag. Compare this to the more elaborate examples shown earlier for the Gempei War period.

Azukizaka in 1564 the *monto* (members) of the Ikko-ikki advanced against Tokugawa Ieyasu with this slogan on the tablets in their helmets. A white flag with this inscription has also survived; it was flown by the Mori navy, and similar ones may also have been used by monkish armies.

In a similar way the Christian army who defended Shimabara in 1638 also made use of their own religious designs on flags, with depictions of Christian crosses, angels and other Christian motifs. One flag from Shimabara (see Plate J8) has miraculously survived, and depicts two angels adoring the Blessed Sacrament. During the siege of Osaka by Tokugawa Ieyasu in 1615 there were many Christian warriors in the garrison, and one commentator noted that on the ramparts 'there were so many Jesu's crosses and Sant'Iago's as to make Ieyasu sick to his stomach'. It may be that images of the Virgin were carried into battle in Japan, but none has survived to provide proof. It is important to note, however, *(continued on page 48)*

The *mamori* (shrine amulet) of Tachibana Muneshige (1562– 1642), who fought in Korea and besieged Otsu Castle in 1600. This is a good example of the use of a religious motif as a *mon*.

LEFT The *sotoba* device of Mogami Yoshiaki (1546–1614), a daimyo from northern Honshu, appears prominently on this section of a screen depicting the siege of Osaka Castle; cf the similar *sotoba* motif of the Ikko-ikki, Plate D14. The other flag bears, unusually, an animal – apparently a white rabbit or hare on a red field.

OPPOSITE Christian devices, including crosses and pictures of the Blessed Sacrament, decorated the flags flown at Hara Castle during the Shimabara Rebellion of 1638, as shown in this scroll from the Watanabe Museum, Tottori. See also Plate J8.

that many cross-shaped designs used in Japan had no Christian significance but were derived from the Chinese character *ju*, either meaning the number '10' or simply representing an abstract design based on a horse's bit. The Shimazu of Satsuma provide the best known example of this. The 'St Andrew's Cross' of Niwa Nagahide (see Plate D3) also had no Christian significance, but was derived from the shape of crossed timbers in a Shinto shrine.

HERALDRY AND THE TOKUGAWA SHOGUNS

By the end of the 16th century the system of Japanese heraldry, although still unwritten, was becoming universally recognised and was remarkably uniform in its execution. Mon now appeared on the sails of daimyos' ships and on the costumes of samurai even when not in the field, such as the times when they were guarding a castle. In such a case the samurai's costume would probably be the *kamishimo*, which combined *hakama* (wide trousers) and *kimono* (robe) with the curious stiffened winged jacket called a *kataginu*. The mon were displayed on the breasts and back of the kataginu, and also appeared on the sleeves of the kimono. The all-purpose *haori* jacket would also carry mon. On a larger scale, huge nobori flags were flown from the towers of castles, and mon still appeared on the maku field curtains.

With the triumph of the Tokugawa family at the battle of Sekigahara in 1600 and the beginning of the peaceful Edo Period after the liquidation of the Toyotomi family at Osaka in 1615, we see the final organisation of

The *aoi* (hollyhock) *mon* of Honda Tadakatsu (1548–1610). This is essentially a reworking of the design used by Tokugawa Ieyasu, whom Honda Tadakatsu served loyally throughout his life.

The use of *mon* on the sails of daimyo's ships during the invasion of Shikoku in 1585; Hideyoshi's *kiri* device leads the way.

a heraldic system combining mon and colour, thus effectively making military uniforms out of samurai costume. The designs of mon, the colours of the coats, and the number of flags were all meticulously laid down in regulations, just as was every aspect of samurai life under the Tokugawa. Mon were also found on e.g. sword scabbards, and used for interior design in castles.

The mon most frequently seen during this time was of course that of the shogunal family. Its adoption by the Tokugawa is dated to the time of Tokugawa Ieyasu's father Hirotada (1526–49). Legend tells us that he was resting in the house of a vassal who presented him with some cakes set upon three large wild-ginger leaves on a round wooden platter. Hirotada took the design suggested by this arrangement and made it into his mon. It is usually referred to as *mitsu aoi* ('three hollyhocks') even though it is actually derived from ginger *(asarum caulescens)*. Tokugawa Ieyasu, in his turn, allowed his faithful retainer Honda Tadakatsu to use a variation on the three leaves design as his mon.

A gorgeous display of heraldry was regularly seen during the practice of 'Alternate Attendance', by which the daimyo resided alternately in the shogun's capital of Edo and in

The *yabane* (arrow fletchings) *mon* of Hattori Hanzo (1541–96), the legendary leader of the *ninja* of Iga, who was one of Tokugawa Ieyasu's most trusted generals.

BELOW **The heraldic display of Furuta Shigekatsu (1561–1600); he was outlived by his father Shigenari, a noted tea master, who took part in the siege of Osaka (1614–15) and probably used the same heraldry. There is an amusing story about Shigenari risking death by climbing over the siege lines at Osaka to retrieve an aesthetically pleasing piece of bamboo. (1) Messengers' *horo* and (2) samurai *sashimono* – black with gold *mon*; (3) three-dimensional standard – silver rain hat; (4) *nobori* – white *mon* on dark blue field.**

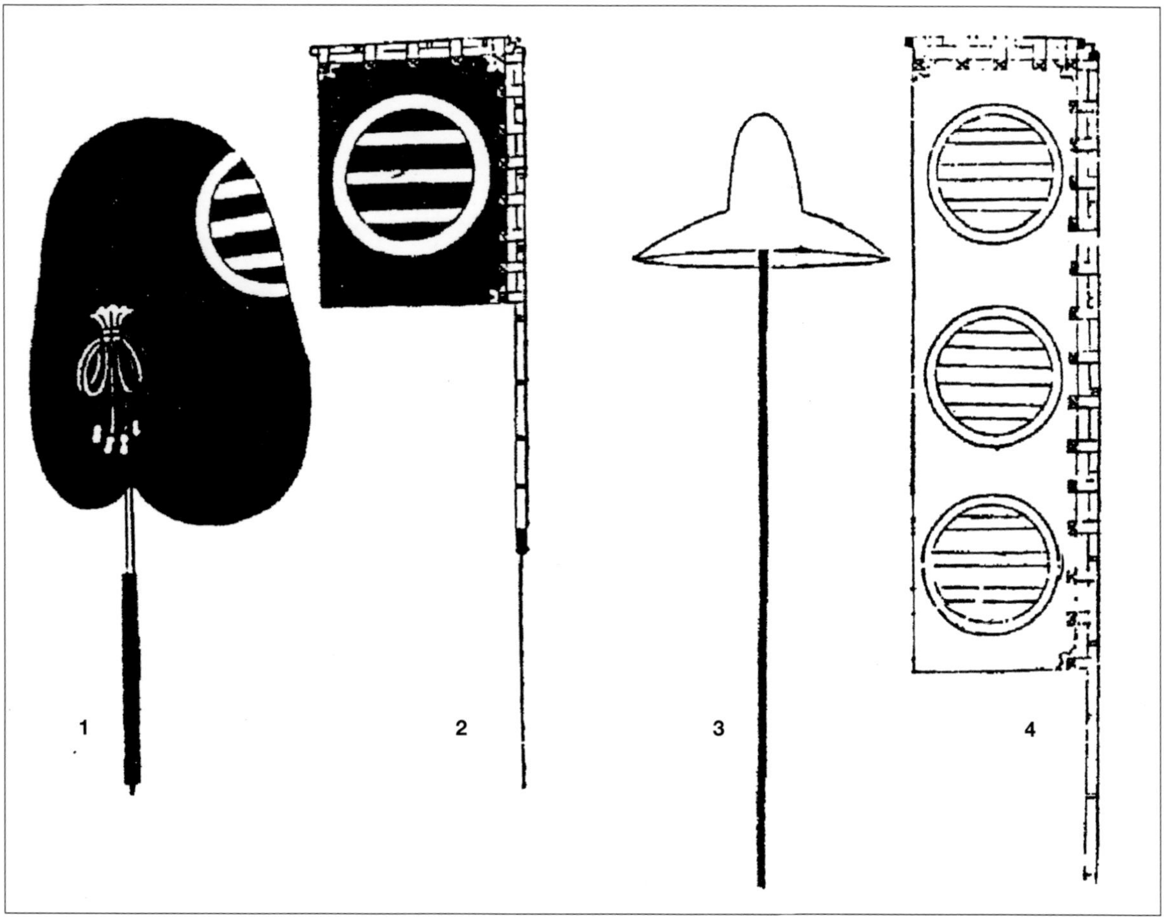

1

2

3

4

their own castle towns. This meant that the roads of Japan were frequently crowded with huge processions passing in both directions; and as it was vitally important to know the relative rank of different families in order to decide who should have precedence, a reference book of daimyos' mon called the *Daimyo Mon Zukushi* began to be published annually. In 1642 the Shogun Tokugawa Iemitsu decreed that all warrior families were to register two mon and never to vary them in any way.

Heraldry and the common people

From the 17th century onwards mon also began to appear as the insignia of people outside the samurai class. A particular boost was given to this practice by the popularity of the *kabuki* theatre. As actors were not

Aoki Shigekane, who served Tokugawa Ieyasu, used this representation of Mount Fuji in white on black on all his flags.

A woodblock print of an actor, showing an example of the *mon* designs adopted by the theatrical profession during the Edo Period. This is from the play *Shibaraku*, and the *mon* is the very unmilitary device of a set of interlocking rice measures.

allowed to use the mon of the actual historical characters they were portraying they created their own, often choosing striking designs that would look most effective on stage. Dynasties of kabuki actors then adopted these mon as their family badges, a practice that was quickly copied by other merchants and tradesmen; and by 1790 the artist Sharaku was able to identify 64 mon of leading actors in Edo.

The growing prosperity of the age encouraged many samurai to adopt new and fanciful mon of their own, and to spend large sums of money having the new designs incorporated into lacquerware, furniture and exterior design. A poem of the time satirised the practice: 'Choosing a mon for your clothes/ Will lead you ultimately to bankruptcy/ With your house and warehouse all lost'.

The flag standard of the Sanada flown during one of the 'Alternate Attendance' processions of the Edo Period when the daimyo visited the capital. The white *mon* on a red field represents six coins, the fee paid to the ferryman across the river of death – see also Plate J1. The three-dimensional standard at right centre is gold above long red tassels.

The sun disc

One flag that was not seen in the streets of Edo at that time, of course, was the *hi no maru* (sun disc), the banner that is now the Japanese national flag. There was no Japanese national device during the age of the samurai, but because Japan (Nihon) was literally 'the Land of the Rising Sun' it is not surprising to see the sun disc design appear among the heraldry of several prominent daimyo families. We noted earlier that Takeda Shingen used a large hi no maru as one of his standards, and this flag is still preserved in the Takeda Museum at the Erin-ji temple near Kofu. Sakai Tadatsugu also used a sun disc uma jirushi, together with nobori that displayed three suns in a vertical row (see Plate D8). Date Masamune is illustrated on a portrait scroll with a sun flag as his personal sashimono.

The most interesting example of the use of the hi no maru, however, is the adventurer Yamada Nagamasa (1578–1633), who is said to have employed it as his banner when he served the king of Siam as his bodyguard. If true, this would make it virtually a Japanese national flag, as Yamada Nagamasa led a sizeable contingent of Japanese mercenaries in Siamese service. Unfortunately, the votive painting presented by Yamada Nagamasa to the Asama Shrine in Shizuoka, which depicts a Siamese ship with him on board, only shows the Yamada family flag of a mon design, and not the hi no maru. The tradition of Yamada using the hi no maru may therefore just be wishful thinking on the part of Japanese patriots at the time when the national flag was introduced in the Meiji Period to fly from ships taking Japanese embassies abroad.

The hereditary principle in heraldry

In 1880 a book was published listing over 500 prominent samurai families with 3,040 mon. About half used mon based on flowers and plant life, a quarter used motifs based on artificial objects, and a quarter used geometric designs. Only a handful have designs based on birds or animals, and none use fish. This is in marked contrast to Western heraldry, where devices of beasts such as lions were always popular. The difference from Chinese and Korean heraldic design is also illustrated by contemporary paintings, where the contrasting colours and bold mon of the Japanese flags stand out very distinctively from the elaborate and complex

OPPOSITE **A three-dimensional standard used by the Sanada during their Alternate Attendance processions – this actual example is preserved in Ueda Castle.**

In Japanese heraldry there are almost no rules for heredity or systems of differentiation between cadet branches comparable to those of European heraldry. This is illustrated here by the Kyogoku family. Tadataka (1593–1637) fought at Osaka, and used this red *nobori* with one Kyogoku *mon* in white. His heir and nephew Tadakazu used a *nobori* with two *mon* in black. See also Plates H6 & H7.

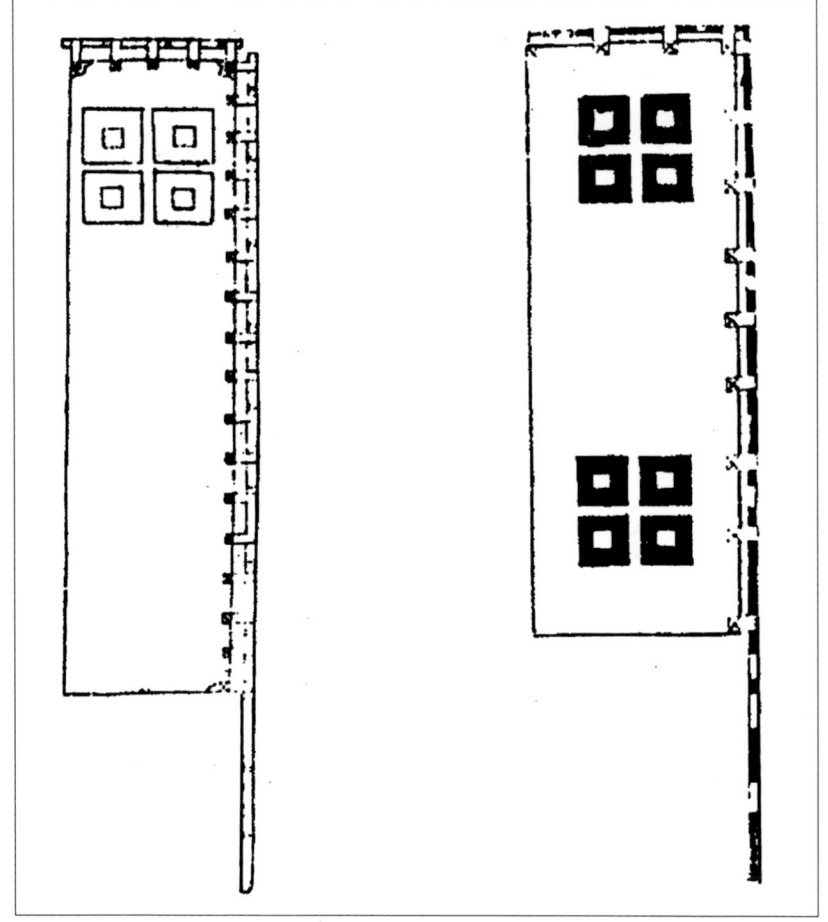

embroidery found on continental flags, which were often of a triangular design. Korean banners also made much more use of animal motifs than did the Japanese.

Although numerous mon existed, and several were used by more than one family, certain examples were very jealously guarded and passed on from father to son like a European heraldic charge. In 1650 a certain Kyuan published a book entitled *O Uma Jirushi* which set out the heraldry of many of the most prominent families of his day. As the then heads of families were the sons or grandsons of daimyo who had fought during the Sengoku period and whose own heraldry is well recorded, Kyuan's book enables us to assess the process involved in passing on a mon or other motif. (Apart from hereditary transfer, mon were sometimes granted to faithful retainers in recognition of services rendered.)

The Japanese hereditary heraldic model is certainly a far simpler system that the European model of successive quartering. In most cases the mon merely passed from one generation to the next unchanged, the only real variation occurring in the design of the field of the flag on which the mon appears – but even this was by no means subject to fixed rules. Only the mon itself, the quintessential element of identification, remains the constant in the colourful and vivid world of samurai heraldry.

The Heraldry of Samurai Armies for Wargamers and Modellers

Wargamers and modellers may be interested by the following notes on how to use the descriptions and illustrations in this book for painting an accurate samurai army in miniature.

As is apparent from the text, there are huge gaps in the records; but the best guide to what a samurai army of the Sengoku period may have looked like are the known designs of *mon* or, in certain cases, the fuller illustrations of heraldic display taken from the *O Uma Jirushi*. I have tried to include as many as possible of these in this book, either in the monochrome illustrations or in the colour plates. As the original source shows 220 different families I have had to be selective. I have therefore omitted those which are unidentifiable, or which are illustrated in my other books. In *The Samurai Sourcebook* (Cassell, 1998), for example, many of the illustrations from the *O Uma Jirushi* are described, but lack of space meant that very few could be illustrated. Of these I have given preference here to any that might be ambiguous from the written descriptions alone.

The great value of the *O Uma Jirushi* is that, although it was not compiled until 1650, its details of flags can probably be safely extrapolated back at least one generation. There is no space here to show the relevant family trees for doing this, and the reader is referred to *The Samurai Sourcebook*, where most of the major families are covered.

The most straightforward armies to create are those of the Gempei Wars with their simple *hata jirushi*. The colourful designs on the armour, produced by making patterns with the lacing of the plates, had no heraldic

The name of Inaba (Ittetsu) Masanari, who died in 1628, crops up in accounts of battles from Anegawa (1570) to Odawara (1590) and the Tokugawa wars. His full heraldic display is shown here: (1) Ashigaru triple back flag – white on dark blue; (2) messengers' *sashimono* – white design on dark blue; (3) samurai *sashimono* – gold; (4) great standard – gold; (5) lesser standard – gold; (6) *nobori* – white on dark blue.

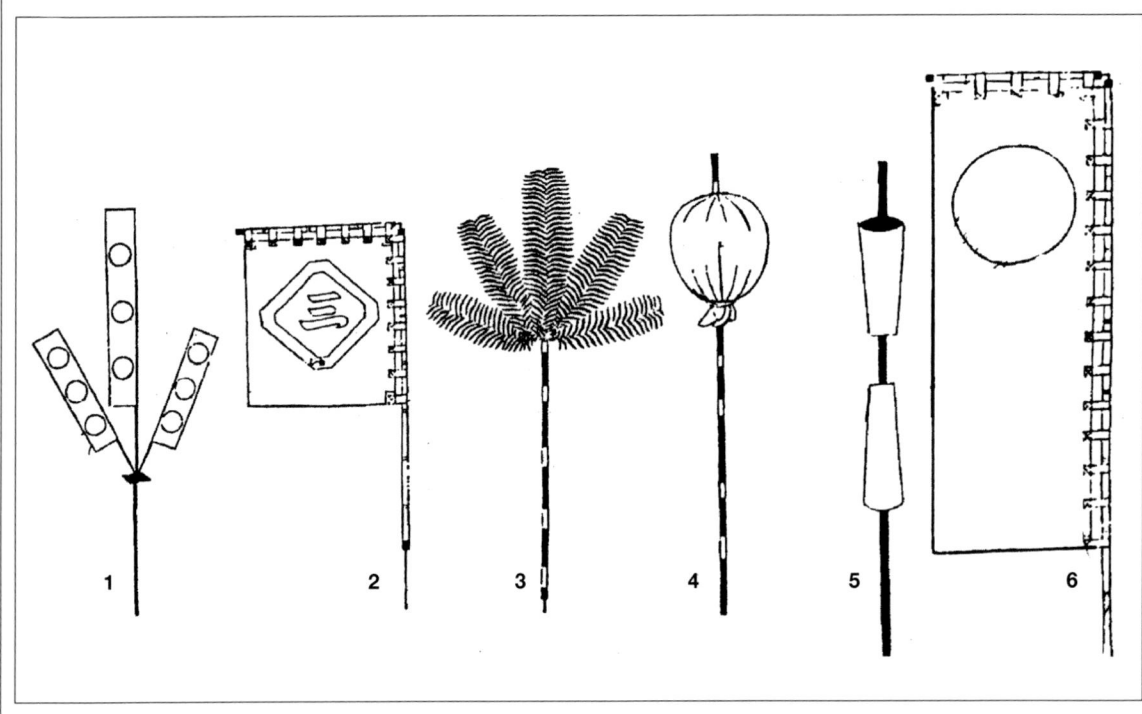

1   2   3   4

significance, although red was supposed to be reserved for generals. With the introduction of the *sode jirushi* and *kasa jirushi* we are able to consider more 'uniform' armies, but the challenge really begins with the Sengoku period.

For a 16th-century headquarters scene, make sure that your daimyo has his *o uma jirushi* and *ko uma jirushi* and any other flags that are known. These are usually noted in the *O Uma Jirushi*, but the drawing is often a bit fanciful, so take heed of my descriptions. There is often a description of a daimyo's helmet too, so make a model ashigaru carrying it, and don't forget the drummers and the conch-blower. The surrounding curtains would bear the daimyo's *mon*.

The *O Uma Jirushi* usually gives additional details about the so-called *ban sashimono*, which may be taken as the design of back flags used by the samurai. These usually bore the *mon*, and in order to create subdivisions within the army it is not unreasonable to place it on different coloured flags. The five colours of red, blue, yellow, black and white were quite common. Complications arose when subordinate units of an army were commanded by generals who used their own designs along with the overall commander's *mon*. The 'Twenty-Four Generals' of Takeda Shingen who were family

Nanbu Toshinao was the son of Nobunao (1546–99), and the heir to one of the great daimyo families of the far north of Japan. (1) Messengers' *horo* – black; (2) samurai *sashimono* – nine gold sun rays; (3) lesser standard – a gold covered basket with a black plume of feathers; (4) great standard – a crane in black on white.

The *fuji* (wisteria) *mon* of Goto Mototsugu, who fought in Korea and was killed at the siege of Osaka in 1615.

members used their own *mon*, but in many cases the distinction between family members, hereditary retainers and allies was a fine one. Thus the Ii fought at Sekigahara under their own banners, not those of the Tokugawa. See my *Ashigaru 1467–1649* (Warrior 29, Osprey) for a more complete discussion of this point.

In most cases the messengers of the *tsukai ban* should be distinguished by a vivid *horo* or *sashimono*. Sometimes a back flag design is also recorded for the common footsoldiers, but the display of the *mon* on the breastplate was practically universal. Usually there would be numerous *nobori*-carriers in the army. If the actual design of a *nobori* is not recorded, then to place the *mon* at the top of a long white flag cannot be far wrong.

For some families active during the early to mid-16th century, or who died out before 1650, often only the *mon* is known; this is particularly true of those families that were destroyed after fighting on the losing side in the famous battles of the Sengoku period. An example is Kasahara Kiyoshige, whose *mon* in shown as Plate C8; his family were wiped out by Takeda Shingen. In such cases the *mon* will have to provide the basis for painting the entire army. Make the *uma jirushi* a large rectangular flag bearing the *mon*. The *nobori* would have either one *mon* at the top, or three in a vertical row.

An authentic-looking army would have numerous *nobori*, and the samurai would display *sashimono* bearing the *mon*. To distinguish units of samurai two historically accurate methods may be adopted. In the case of a family such as the Kasahara, where the base colour appears to have been important, different units may be distinguished by using single, double or triple bands of black or white at the top of the flag. Otherwise, different coloured fields in the five lucky colours can be used. Give the ashigaru the *mon* in gold lacquer on the breastplates and *jingasa* helmets.

Tozawa Masamori (1585–1648) wisely sided with the Tokugawa in 1600. (1) Messengers' *sashimono* – red disc on blue with a white plume. The ashigaru and samurai *sashimono* (not illustrated) were similar but smaller flags, without the plume. (2) Lesser standard – the same coloured flag, suspended from gold antlers; (3) great standard – three gold umbrellas under a black feather plume; (4) *nobori* – black and white.

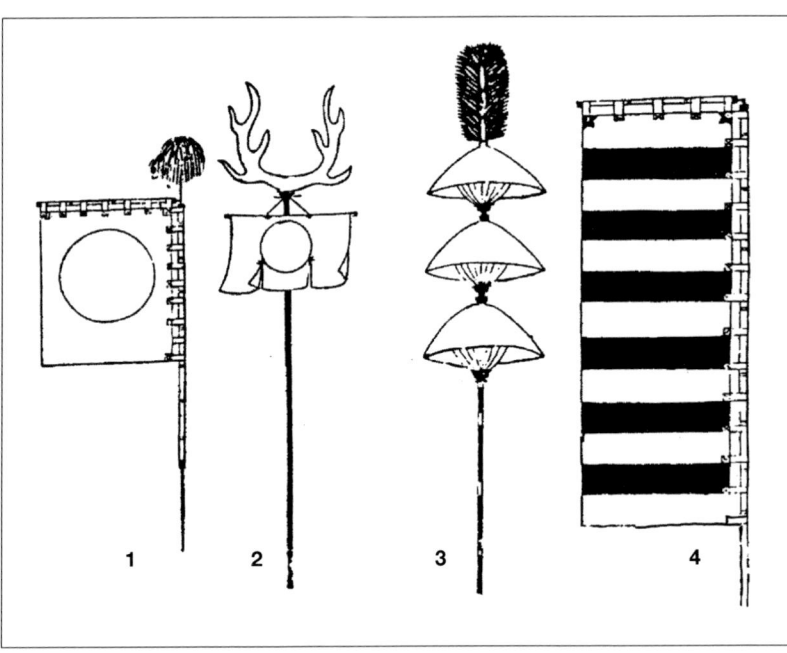

THE PLATES

The demands of space and composition make it impossible to render the separate images of flags and standards on these plates to a constant scale. Their approximate relative sizes may be inferred from desriptions of various types of flags in the text and from some of the monochrome illustrations.

A: THE BATTLE OF DAN NO URA, 1185

The battle of Dan no Ura, which ended the Gempei Wars, was fought at sea; and the legends say that the waves turned red with the blood of the slain and the red dye from the flags of the defeated Taira family, while the white flags of the Minamoto were flaunted in triumph. We see here the butterfly *mon* of the Taira **(A4)** and the gentian of the Minamoto **(A3)** on flags of *hata jirushi* form. The main subject, however, recalls the little known contingent of soldier-priests from the Kumano Shrine, who fought under the white banner of Kongodoji, guardian of the three shrines of Kumano **(A1)**. The Kumano standard-bearer is approached by a comrade who brings him a red silk banner **(A2)** bearing the imperial chrysanthemum of the child emperor Antoku, who was drowned during the battle. This flag was flown from several ships that were not carrying the emperor, so as to confuse the Minamoto. The three-legged hawk emblem **(A5)** was another device used by the Kumano Shrine forces.

B: THE MUROMACHI PERIOD, 1336–1568

This plate depicts some famous flags used by the Kusunoki family and their opponents during the Nanbokucho Wars; other flags take the story on into the Muromachi Period, which ended in 1568 with the abdication of the last Ashikaga shogun.

B1: The 'chrysanthemum on the water' flag of Kusunoki Masashige (1294–1336), which symbolised his loyalty to the emperor in a unique way.

B2: The same design but with the addition of the Chinese slogan described in the body text on page 13.

B3: Ashikaga Shigeuji (1434–97) was styled Koga-kubo, 'governor at Koga' in the war-torn Kanto region. His banner bore both the Ashikaga family's *kiri* emblem and a *hi no maru* sun disc. This is one of the earliest depictions on a samurai's banner of the motif that was later to appear on Japan's national flag.

B4: The flag of the last Ashikaga shogun, Yoshiaki (1537–97), who combined a red sun disc with the ancient Ashikaga device of an evocation of the god Hachiman. The *hi no maru* is correctly called a 'rising' sun, although in the West that term is nowadays used specifically for the rayed sun battle flag used during World War II.

B5: Ouchi Yoshitaka (1507–51) was typical of the old aristocracy who were overthrown by their own vassals during the turmoil of the *gekokujo* following the Onin Wars. His flag bears his *mon* below the titles of the *kami* Myomi Dai Bosatsu, Hachiman Dai Bosatsu, and the deified Sugawara Michizane. See the illustration on page 17 for a more elaborate version.

The other illustrations show examples of the heraldic innovation of the 14th–15th centuries – the use of a small flag attached to the samurai's person, rather than a separate banner carried by a servant.

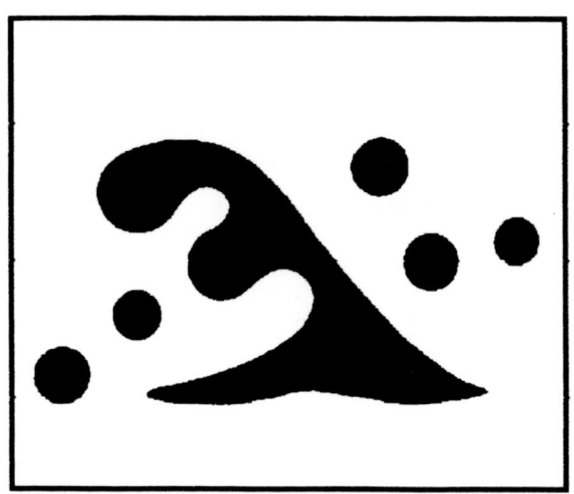

The wave *mon* of Saito Dosan (1494– 1556), an alternative design to the device shown as Plate C9.

The standard of Akechi Mitsuhide, the murderer of Oda Nobunaga in 1582; compare this with his *nobori* shown as Plate D10. The *nobori* was often augmented with a smaller flag at the top of the shaft, usually in the form of a *hata jirushi* streamer. Designs varied, but they rarely duplicated the overall design of the *nobori*.

B6: A *kasa jirushi* attached to a ring at the rear of the helmet.
B7: A *kasa jirushi* attached to the front crest of a helmet.
B8: A simple *sode jirushi*.
B9: A more elaborate *sode jirushi* which bears an invocation of various Shinto deities.

C: THE *SENGOKU DAIMYO*, 1467–1560

The rival regional warlords of the Sengoku Jidai continued to use the *hata jirushi* type of flag, but were also the first to use the *nobori* and *sashimono*. This selection of flags shows the development of the styles.

C1: The 'old-fashioned' *hata jirushi* of Hojo Soun (1432–1519), the upstart founder of the Odawara Hojo family, who appropriated both the ancient Kamakura Hojo's name and their *mon* of three triangles arranged in a triangle.

C2: A few decades later Hojo Ujiyasu (1515–70), the grandson of Hojo Soun, used this large *nobori*-shaped flag in the five 'lucky colours' as his standard.

C3 & C4: Uesugi Kenshin (1530–78) used these two standards, a large flag and a fan of lacquered wood, both bearing the *hi no maru*. (*Erratum:* the ground of the flag should be dark blue, not black.)

C5: Imagawa Yoshimoto (1519–60), who was killed at the battle of Okehazama in 1560, used this design of a comb on his standard.

C6: The flag flown by the warriors of Suwa Yorishige (1517–42), a deadly enemy of Takeda Shingen. The *nobori*, *sashimono* back-flags and large standard used by this army would have differed only in their dimensions.

C7: Yamamoto Kansuke (died 1561), a loyal retainer of Takeda Shingen, used the design of a *vajra* on his followers' *sashimono*. Alternative versions used white sections above black, with the *vajra* in reversed colours.

C8: The *kikyo* or Chinese bellflower used on all flags by Kasahara Kiyoshige, who was defeated at Shika by the Takeda army in 1547. In this and C9, the other clan flags may be inferred from this basic design.

C9: The *nadeshiko* or carnation used on all flags by Saito Dosan (1494–1556), shown here on a *sashimono*. Saito used an additional *mon* of a wave, illustrated on page 57.

D: THE AGE OF ODA NOBUNAGA, 1560–82

A selection from the heraldry used by Oda Nobunaga and by his contemporaries. (Flags painted as viewed from reverse side, but any ideograms not reversed.)

Oda (Kambe) Nobutaka (1558–83), the son of Oda Nobunaga, sits proudly in front of his standard in the classic position of the samurai general; the umbrella is gold, the *hata jirushi* red and white. Nobutaka committed suicide when Gifu Castle fell to Toyotomi Hideyoshi in 1583.

1

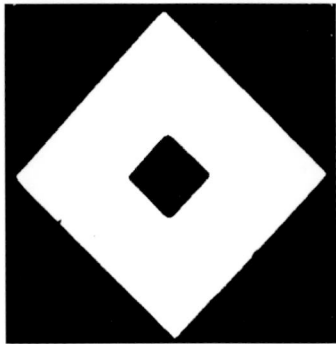

2

3

4

D1 & D2: The three-dimensional red umbrella standard, and coin design *nobori,* of Oda Nobunaga (1534–1582).

D3: Niwa Nagahide (1535–85) fought at the battle of Anegawa (1570) between Nobunaga and the forces of the Asai and Asakura. His standard bore a red saltire.

D4: Niwa Nagahide's *nobori* and *sashimono* were black and white.

D5: The standard of Okudaira Nobumasa (1555–1615), who defended Nagashino Castle during the famous siege of 1575, bore his red fan *mon* on his standard.

D6 & D7: Okudaira Nobumasa's *nobori* and *sashimono* kept the same colours but were of simpler design. His messengers wore *horo* striped in red and white.

D8: The *nobori* of Sakai Tadatsugu (1527–96), who fought at Nagashino.

D9: Sakai Tadatsugu's samurai and messengers wore *sashimono* with one sun disc.

D10: The light blue *nobori* with triple white *mon* of Akechi Mitsuhide (1526–82), who was defeated at Yamazaki by Hideyoshi Toyotomi after assassinating Oda Nobunaga.

D11 & D12: The fan standard and striped *nobori* of the Christian daimyo Takayama Ukon (1553–1615), who fought at the battles of Yamazaki and Shizugatake (1583) during Hideyoshi's consolidation of power after Oda Nobunaga's death.

D13: The 'sword of Fudo' was displayed on the *nobori* of Ikeda Tsuneoki (1536–84), who fought at Okehazama and was killed at Nagakute.

D14: A banner of the Ikko-ikki of the Zempukuji, who defied Oda Nobunaga during the long siege of Ishiyama Honganji. It shows the Buddhist device of a gold *sotoba* on red. The flag is fully reversed.

E: THE AGE OF TOYOTOMI HIDEYOSHI, 1582–98

Toyotomi Hideyoshi, the successor of Oda Nobunaga, took into his service many of Nobunaga's followers. (Flags shown from reverse side but ideograms not reversed.)

E1 & E2: The original version of the gourd standard, and the golden *nobori* of Toyotomi Hideyoshi (1536–98). The standard was later elaborated into the 'thousand gourd' version, of which a reconstruction is illustrated on page 24.

E3: The purple and white *nobori* of Katagiri Katsumoto, one

LEFT

(1) The *takaha* (hawk feathers) *mon* of Asano Nagamasa (1546–1610), the brother-in-law of Toyotomi Hideyoshi, whom he served in Korea.

(2) The *kuginuki* (tool) *mon* of Hori Hidemasa (1553–90), a loyal general of Hideyoshi who fought at Yamazaki and Shizugatake.

(3) The *zen* (coin) of Sengoku Hidehisa (1551–1614), another of Hideyoshi's generals. A similar *mon* appears on the *nobori* of Oda Nobunaga in Plate D2.

(4) The wheel *mon* of Sakakibara Yasumasa (1548–1606), one of the most trusted followers of Tokugawa Ieyasu. This is variously interpreted as the Buddhist 'wheel of the law' (cf Plate E7), or just as a simple cartwheel.

of the 'Seven Spears of Shizugatake' – the valiant warriors who distinguished themselves at the famous battle in 1583.

E4: Ota Kazuyoshi, who fought in Korea, used the first character of his name on his standard.

E5, E6 & E7: Ikoma Chikamasa (died 1598) fought in Korea. All his devices except the lesser standard made use of the motifs of stars (great standard, E6), and the 'wheel of the law' (*sashimono*, E5, and *nobori*, E7).

E8: The *nobori* of Ukita Hideie (died 1662), who was Hideyoshi's commander-in-chief in Korea in 1592–93.

E9 & E10: The *nobori* and messengers' *horo* of Todo Takatora (1556–1630), who was appointed an admiral in the Korean campaign, and went on to serve the Tokugawa. They provide an example of differing colours within one daimyo's heraldry; and of a *horo* combined with an additional device worn on the warrior's back like a *sashimono*.

E11: The large *sashimono* of the samurai of Ishida Mitsunari, who fought in Korea, and who later led the army defeated by Tokugawa Ieyasu at the battle of Sekigahara (1600). As was almost invariable, he paid for defeat with his life.

E12: The three-dimensional standard of Kuki Yoshitaka (1541–1600), who served both Nobunaga and Hideyoshi as a naval commander.

E13: Kuki Yoshitaka's *nobori*.

E14: Kuki Yoshitaka's alternative *nobori* bearing the Chinese characters *a ra ha*.

The triple *tomoe* design – see Plate F4 – was used by several samurai, including Kobayakawa Takakage (1532–96), the hero of the battle of Pyokje, Korea, in 1593.

Kato Kiyomasa in action in Korea in 1592, showing various heraldic devices – see Plate G. The 'Nichiren' slogan is shown here on the lower panel of a *nobori* carried on a footsoldier's back; this is a variation on Kato's better known *hata jirushi* version (Plate G1). Another footsoldier carries his lesser standard (Plate G2), which is shown as a *nobori* with the famous blue 'snake's eye' *mon* under a fan of paper feathers. The *mon* also appears on the helmets of his ashigaru. Kato Kiyomasa is wearing his famous silver helmet with the sun disc motif – see page 29.

F: MAEDA TOSHIIE AT THE SIEGE OF SUEMORI, 1584

The relief of Suemori Castle in 1584 by the army of Maeda Toshiie (1538–99) was one of the most celebrated actions fought by this famous general, who served Hideyoshi from his castle town of Kanazawa. This plate, which is based on a painted scroll owned by the Maeda family and never published outside Japan, gives an impression of the Maeda hierarchy in a typical heraldic ensemble of the period.

Maeda Toshiie **(F1)** and his son Toshinaga (1562–1614) **(F2)**, who are sitting on camp stools, have their spectacular helmets **(F3A, F3B)** displayed on spears carried by their ashigaru attendants as if they were personal flags; the helmets are hugely extended with lightweight materials into the shape of courtier's caps, gilded and silvered respectively, with added white and black horsehair fringes. Behind Toshiie flies the banner of Shoki, the queller of demons **(F5)**, which he used as his o uma jirushi ('great standard'); a subsidiary standard was a three-dimensional shape with a feather plume ball **(F6)**. The scroll shows two other flags: **F4** has the 'three commas' motif in white on red, and **F8** the characters for 'gold' and 'child' in gold on blue. A mass of the usual nobori are white with the Maeda mon in black. This also appears on the soldiers' helmets and breastplates in gold.

G: KATO KIYOMASA AT CHINJU, 1593

The ultimate samurai achievement was to be seen to be the first into battle. When the objective was a castle, the proof of entry was to have one's flags flown from the wall; and on occasion standard-bearers are known to have thrown their lord's flag over the wall for the samurai to follow. During the Japanese attack on Chinju in Korea in 1593, the retainers of Kato Kiyomasa (1562–1611) and Kuroda Nagamasa (1568–1623) were in competition to be the first through the breach. Kiyomasa's standard-bearer Iida Kakubei throws the 'Nichiren flag' **(G1)** over the wall – an ancient hata jirushi with the Buddhist inscription 'Hail to the Lotus of the Divine Law'. Iida was one of Kato's three most loyal retainers, who took it in turns to carry their lord's flag; Iida wore as a sashimono a plume of white feathers, his comrade Morimoto Gendayu a similar display of black feathers, while Shobayashi Shunjin wore one of white over black.

Another flag-bearer **(G2)** follows Iida with Kato's ko uma jirushi ('lesser standard'), which was of a multiple design – a nobori bearing Kato's circular blue 'snake's eye' mon surmounted by a three-dimensional fan device of white paper strips fluttering from 'fronds' mounted on a golden circlet. To the rear is the Kuroda standard-bearer **(G3)**, frustrated in his attempt at glory; the purple nobori bears Kuroda's mon in white, but the small sashimono has a black disc on white.

H: THE AGE OF TOKUGAWA IEYASU, 1598–1615

Tokugawa Ieyasu (1542–1615) and his son Hidetada (1579–1632) consolidated their family's position through victories at Sekigahara and in subsidiary campaigns in 1600, and by their final crushing of Hideyoshi's son Hideyori at the winter siege and summer battle of Osaka in 1614–15.
H1 & H2: The famous golden fan and 'boar's eye' standards of Tokugawa Ieyasu.
H3 & H4: The flag great standard, and three-dimensional 'fly trap' lesser standard, of Tokugawa's leading retainer Ii

An example of the *shakujo*, the rattle used by wandering *yamabushi* of the mountain cult, reproduced in very large form as a standard; cf the huge golden Tsugaru version in Plate I.

Naomasa (1561–1602), whose 'red devils' fought at Sekigahara (1600). Red was the predominant colour in all their heraldry.
H5: The *nobori* of Naoe Kanetsugu (1570–1619), who fought against Tokugawa forces at Hasedo in 1600 in the service of Ishida's ally Uesugi Kagekatsu.
H6 & H7: The *nobori* and *sashimono* of the Tokugawa retainer Kyogoku Tadatsugu (1560–1609), who defended Otsu Castle prior to the battle of Sekigahara.
H8 & H9: Fans appear on the *nobori* and *sashimono* of

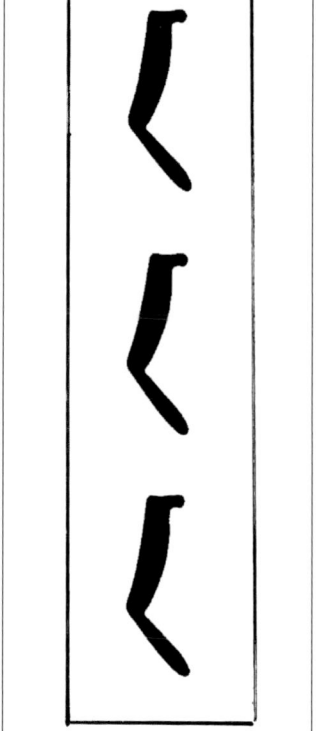

Satake Yoshinobu (1570–1633) as a rigid 'cut-out'; he took a prominent role in the battle of Imafuku during the Osaka Winter Campaign of 1614–15.

H10–H13: The striking and varied heraldic designs used by Hoshina Masamitsu (1561–1631), who fought at Osaka in 1615. The *sashimono* **(H10)** and *nobori* **(H13)** show the *mon*; this was repeated in triple form, one block above two, on small flags protruding from the top of his messengers' all-red *horo*. His great standard **(H12)** was a spray of peacock's feathers, his lesser standard **(H11)** a three-dimensional gold flower shape.

H14: Mukai Tadakatsu, who acted as Tokugawa Ieyasu's admiral and fought at Osaka, used a flag with the character *mu* as his personal *sashimono*. His great standard **(H15)** was a three-dimensional golden bell.

H16: Makino Tadanari, who fought at Osaka, used this three-leaved flower on his standard; but his other flags displayed a ladder design. The *sashimono* had ten rungs **(H17)**; those of his messengers bore this motif in white on a red ground. His *nobori* are illustrated with yellow seven-runged ladders on a black ground.

I: TSUGARU NOBUHIRA AT HIROSAKI CASTLE, 1610

The Tsugaru family became dominant in the far north of Japan, a position they strengthened by their support for the Tokugawa family at the time of Sekigahara. In this plate Tsugaru Nobuhira (1586–1631) marches proudly through the snow to take possession of his newly built castle of Hirosaki. The most prominent motif is his *mon* of a swastika, which appears in red on the *jingasa* helmets and breastplates of the lowly ashigaru, and in gold upon the red *sashimono* of the samurai (I4, I5). Red swastikas also appear on the numerous white *nobori* flags (I2). The *o uma jirushi* (I1) uses different religious symbolism: a huge, gold-lacquered, three-dimensional *shakujo*, the metal-ringed rattle used by the itinerant members of a mountain-based Buddhist sect to drive wild animals away during their pilgrimages through the high country. The heavy standard is socketed and tied to the back of a strong soldier and steadied by extra ropes; note the double red *sashimono* of the rope men. Carried behind Tsugaru, wearing his characteristic antlered helmet, is his lesser standard (I3), a flag bearing a gold disc on white.

J: THE LAST REBELS: OSAKA, 1615, & SHIMABARA, 1638

A selection of heraldic devices displayed by the Toyotomi forces during the Winter and Summer Campaigns of Osaka, and by both sides in the bitter Shimabara Rebellion of 1638 – the final uprising against the Tokugawa hegemony.

J1: Sanada Yukimura (1570–1615) was a defensive expert whose skilful command of Ueda Castle in 1600 prevented Tokugawa Hidetata from joining his father for the battle of Sekigahara; Sanada was later the fortifier and commander of

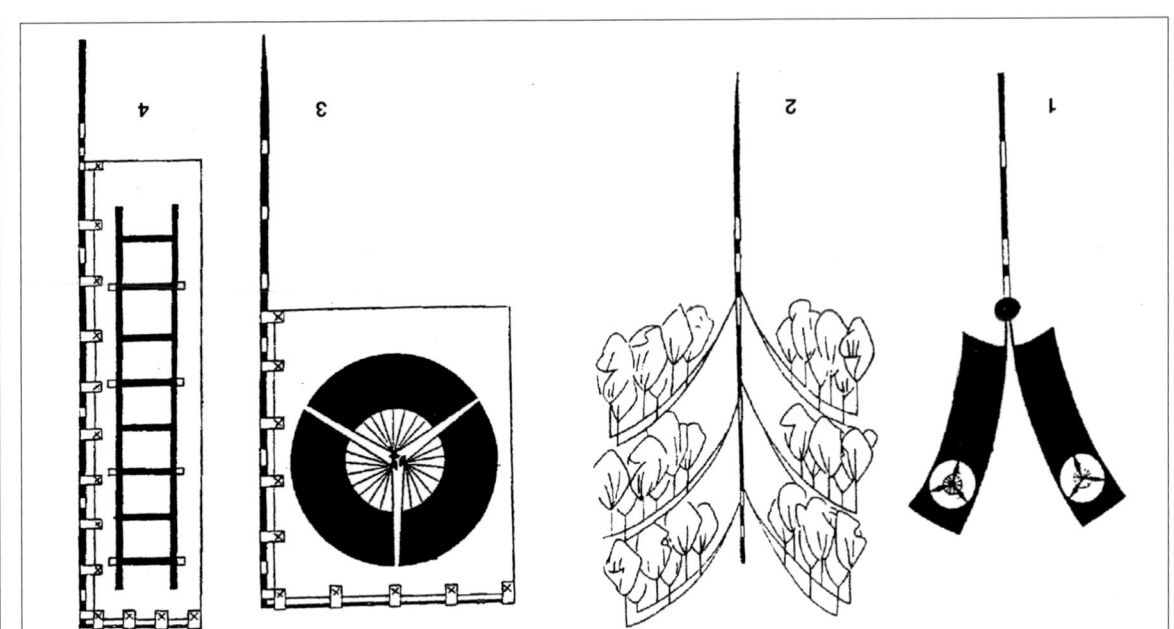

ABOVE Ogasawara Tadazane was the son of Ogasawara Hidemasa, whose heraldry appears in Plate J. Tadazane's lesser standard (1) consists of white paper shapes rather like *origami*, suspended from a 'tree'. The great standard (2) has the Ogasawara *mon* in white on dark blue, while the *nobori* (3) is black and white. The samurai *sashimono*, not illustrated, was a smaller version of the *nobori*.

(1) Samurai *sashimono* – double flags, white on black; (2) lesser standard – a white paper tree; (3) great standard – black on white; *nobori* (4) – black ladder on white.

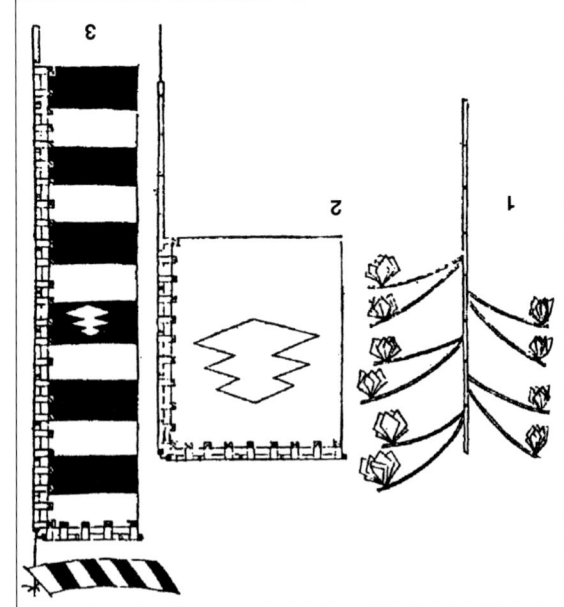

J2: Ono Harunaga (died 1615), another of the defenders of Osaka Castle in 1614-15. His samurai wore this *sashimono* with the 'six coins' motif, as also seen on the flag standard illustrated on page 51.

Osaka, used a hat design on his *nobori* and *sashimono*. His son's *nobori* is illustrated on page 62.

J3–J5: The heraldry of Ogasawara Hidemasa (1569-1615), who served the Tokugawa; his son Tadazane fought at Shimabara. His *nobori* (J5) and small *ashigaru* back-flag (J4) bore his *mon* in white on a red ground; his great standard in Shimabara, displayed a single large *mon*. His samurai wore square black-and-white striped *sashimono* as at J3 (erratum – incorrectly presented here as a rope-braced standard). His messengers' *horo* are illustrated as black, patterned all over with small white spots in a 'starry sky' effect (cf.Plate E6).

J6: Doi Toshikatsu (1573-1644), whose 'yellow regiment' served the Tokugawa, used this very distinctive 'water wheel' motif on his *nobori*.

J7: The *nobori* of Matsukura Shigemasa (1574-1630), whose persecutions sparked the Shimabara Rebellion.

J8: The Christian Hara flag, hand-painted by one of the defenders of Hara Castle and miraculously preserved to this day.

J9: The *nobori* of Itakura Shigemasa (1588-1638), who was killed at Hara in 1638. His samurai wore white *horo* and his messengers black.

J10: Itakura Shigemasa used this three-dimensional standard of a white Chinese paper lantern surmounted by a large black tassel.

J11 & J12: Great standard – a three-dimensional cylinder – and *nobori* of Matsushita Shigetsuna (1580-1628), whose father had served Hideyoshi.

BELOW Matsudaira (Okochi) Nobutsuna (1596-1662), a kinsman of the Tokugawa, succeeded Itakura Shigemasa as commander of the shogun's forces against the Shimabara rebels in 1638, and brought about the fall of Hara Castle.
(1) Samurai *sashimono* – double flags, white on black; (2) lesser standard – a white paper tree; (3) great standard – black on white; *nobori* (4) – black ladder on white.

INDEX

Figures in **bold** refer to illustrations
Flags, *horo*, *mon*, and standards for named individuals and families are entered under their names